FURROWS IN THE SKY
The Adventures of Gerry Andrews

FURROWS IN THE SKY

THE ADVENTURES OF
GERRY ANDREWS

JAY SHERWOOD

ROYAL **BC** MUSEUM
PUBLISHING
Victoria, Canada

Edited, produced and typeset (in Bembo Std 11/13) by Gerry Truscott, with
    editorial assistance from Alex Van Tol.
Digital imaging by Carlo Moçellin and Gerry Truscott.
Cover design by Jenny McCleery.
Index by Carol Hamill.
Printed in Canada by Friesens.

Library and Archives Canada Cataloguing in Publication

Sherwood, Jay, 1947-
    Furrows in the sky : the adventures of Gerry Andrews / Jay Sherwood.

Includes bibliographical references and index.
Issued also in electronic format.
ISBN 978-0-7726-6522-5

    1. Andrews, Gerry, 1903-2005. 2. Surveyors – British Columbia –
Biography. 3. Teachers – British Columbia – Biography. 4. Soldiers –
British Columbia – Biography. 5. Photographic surveying – British
Columbia – History – 20th century. 6. Forest surveys – British
Columbia – History – 20th century. 7. British Columbia – Biography.
I. Royal BC Museum II. Title.

TA533.A54S54 2012   526.9092   C2012-900424-3

Contents

To Gerry's mentors,
F.D. Mulholland, Frank Swannell,
Ellwood Wilson and Fred Haggman,
and to his life-long friends,
Lorne Swannell, Bill Hall, Art Swannell and Jack Aye.

Preface

One afternoon in late August 1983, when I and my family lived in Vander-hoof, there was a knock on the back door. I opened it and a small, neatly dressed man with white hair introduced himself. When I had become interested in writing a book about the famous BC land surveyor, Frank Swannell, almost everyone who knew about him said that I should contact Gerry Smedley Andrews, former surveyor general of British Columbia and a close friend of Swannell. Gerry and I had corresponded a few times. He was returning from his summer cabin in Atlin to Victoria and wanted to meet me. I was thrilled that someone who had such a prominent role in the BC government and the province's history would want to visit, and we spent an enjoyable two hours in conversation. (I have since learned that Gerry's interest in people and his ability to converse with almost anyone was one of the character traits that endeared him to so many.)

My next meeting with him was in Victoria during the Christmas season of 1986. One evening I interviewed Gerry Andrews and Art Swannell, Frank's son, at Art's house, and Gerry invited me to stop by his house the next day. This was my first visit to Andrews' well-known office, a room filled with books and pictures, a cabinet of files and the aromatic smell of pipe tobacco. He provided more information about Frank Swannell and pulled out books and files of information that he thought would be useful for me. I was impressed by his wealth of knowledge and willingness to share it, and how he carried on such a detailed and interesting conversation so easily.

During the following years I met with Gerry Andrews a few more times. But the 1990s was a difficult time to get history books published in British Columbia, and I was busy teaching and helping raise a family, so I shelved the book on Frank Swannell and lost contact with Andrews. When I was finally able to get *Surveying Northern British Columbia*, the first book on Frank Swannell's surveying career, published in 2004, Andrews was one of

Gerry Andrews in 1930 at Monument 259, a Canada-USA border marker on Mount Hefty near the southeastern corner of BC.

the two non-relatives in my dedication. I was amazed and pleased to find that Andrews was still alive at 100 years, and it was a great pleasure to meet him again and personally give him a copy of the book. His daughters, Mary and Kris, read the book to their father, and they told me that he listened attentively.

I am very pleased that the Royal BC Museum is publishing this book in conjunction with the centennial of the BC Forest Service and the arrival of Gerry Andrews' material in the BC Archives. Andrews played an important role in BC's history during a long and adventurous career that included being a rural school teacher, a forester, a soldier in World War II and a surveyor. He revelled in the places he saw and the people he met, and in the process made many life-long friends. Like his mentor and friend, Frank Swannell, Andrews travelled extensively around the province in his work. He was interested in BC's history, particularly as it related to his work, and he took many photographs of the locations he visited. Unlike Swannell, Andrews did not keep a regular diary, but during his retirement years, he wrote extensively about his many adventures. For each article he usually included historical research for context, and he would visit the area again and contact people he knew during that experience. He published several of his articles in the magazines of the BC, Alberta and Manitoba historical societies, and in *Link*, the Association of BC Land Surveyors' journal. He also self-published *Metis Outpost*, a book describing his two years of teaching in the Peace River area and the two summer horseback trips he took while living at Kelly Lake. He gave many public presentations, including a few for the Royal BC Museum, and completed some unpublished manuscripts. Andrews wrote and travelled throughout BC until his early 90s, spending many summers at his cabin in Atlin.

In this book I attempt to combine into one narrative everything I could find that was written by and about Gerry Andrews. I refer extensively to his

A Fairchild FC2W in flight over Camp Number 8, a logging area near Nimpkish Lake. G-CYXN was one of five Fairchilds that the RCAF purchased in 1928 to use for aerial photography. This photograph was taken en route to Andrews' 1934 forestry work. (BC Archives I-68259.)

many articles and manuscripts. Thanks to the Andrews family, particularly his daughter, Mary, I have had access to Gerry's correspondence, unpublished manuscripts and diaries. I have used these to provide additional information and insights. In addition, I have interviewed people who knew him.

Gerry Andrews' particular contribution to our province's history was his work in aerial photography. In the late 1920s and early 1930s this was an important new world-wide technology and Andrews, working for the BC government's Forest Branch, was the leader in developing and implementing aerial photography in British Columbia. He headed the new generation of foresters who took to the air and developed quicker and more accurate methods of inventorying, mapping and monitoring the province's vast, extensive forests. In the process Andrews conversed with leading authorities on aerial photography around the world, and he was the first person in the province to spend extensive time in England and Germany working with the top people in the field. Lorne Swannell, chief forester for the BC Forest

Service from 1965 to 1973, believed that aerial photography was the main technology that revolutionized forestry in BC.

Andrews applied his expertise in aerial photography with the Canadian army during World War II. He commanded a section that used aerial photographs to map the Normandy coastline, enabling the military leaders of the Allies to select the beaches that they would use for the D-Day landings.

After the war, he became head of the Aerial Survey Division for the government's Survey Branch. He coordinated the aerial photography of the entire province and modernized the government's mapping techniques. He also brought accessibility and awareness of the value of aerial photographs to the general public. Andrews finished his career as the province's surveyor general for 17 years. Under his leadership the Surveys and Mapping Branch played an important role in most of the large government and private projects that occurred during British Columbia's economic boom in the 1950s and 1960s.

Gerry Andrews was known for his innovative ideas, his camaraderie with his employees and his travels throughout the province, visiting surveyors in the field with his red flannel sock containing a bottle of hooch.

Early Adventures

"Mid December 1903 in Winnipeg's deep freeze, at 124 Colony Street, my first sniff of air was the rich aroma from Shea's brewery across the road." This was how Gerry Smedley Andrews described his birth. He was the fourth child and oldest surviving son (an older brother had died in infancy) of Thornton and Emma Andrews. Thornton was a pharmacist who came to Winnipeg to open his own drug store shortly before Gerry's birth.

A special childhood memory for Gerry was seeing an uncommon cosmic event. The upstairs bathroom window "commanded an excellent view of the western sky. In May 1910, before my seventh birthday, Mother with firm deliberation took me up there one evening to see Halley's Comet, then conspicuous in the twilight sky after sunset, its spectacular tail fanning out vertically into the darkening firmament above."

Tragedy struck the family in 1913 when Emma Andrews died. Thornton remarried two years later and had two more children by his second wife, but the marriage created conflict in the family.

During World War I all of the provinces passed legislation prohibiting the consumption of liquor. In Manitoba this was done in June 1916. The only legal way to obtain liquor was a doctor's prescription for medicinal alcohol. Thornton Andrews' strong personal and religious beliefs led him to refuse carrying alcohol in his drug store. His business declined and he eventually declared bankruptcy. To help provide the family with some income Gerry worked after school and during the summer as a messenger boy and office boy, and he also had a paper route.

By the spring of 1918, almost four years into World War I, many farms in Canada were struggling to maintain their productivity, because so many men were in service or had become casualties. The Perth *Courier* described the situation in an article titled, "Lack of Food Threatens the Battle Line":

The only thing that balks German ambition is the battle line in France ...

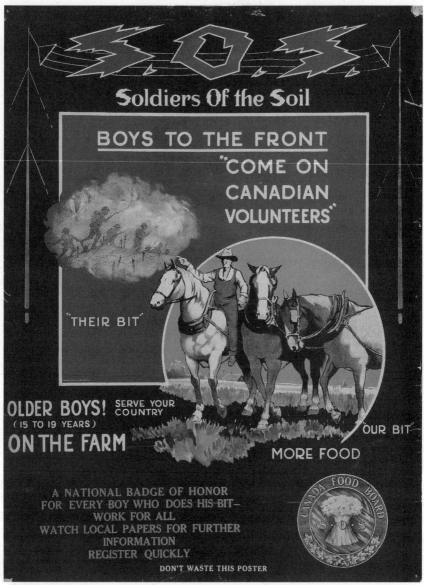

Soldiers of the Soil recruiting poster. (Canadian War Museum 19900076–819.)

and the British navy. The only thing that sustains our men on land and sea is food. One year ago only the enemy was on rations. Today Great Britain, France and Italy are on rations. Today Germany controls the wheat lands of Roumania, Russia, Poland and Ukraine. Today the shadows of hunger, famine, disease and death hang over the Allies. Upon the 1918 crop from Canada and the US depends the fate of the democratic people of the world.

To ensure that enough people worked on the farms, the Canadian government established a Soldiers of the Soil program (the United States implemented a similar one). The Canadian War Museum describes the program:

> Farm labour shortages led the authorities to ask older children and adolescents for help. Soldiers of the Soil (SOS) was a national initiative run by the Canadian Food Board. It encouraged adolescent boys to volunteer for farm service and recruited 22,385 young men across the country. Many came from urban high schools and lived on rural farms for terms of three months or more. In exchange for their labour, SOS recruits received room and board, spending money, and – in the case of high school students – exemption from classes and final exams. On completion of their term and "honourable discharge" they also received an SOS badge acknowledging their service, often at a community ceremony.

Soldiers of the Soil were supposed to be at least 15 years old, but when officials came to Andrews' school in April 1918 he signed up at 14 with his parents' approval. Andrews would be able to make his contribution to the war, bring some income to the family, and be one less person to feed at home. It also presented him with an opportunity to get away from the family difficulties and be independent.

Andrews recalled his experience in a later article, "Reminiscences of a Soldier of the Soil":

> I was sent to a wheat farmer at Purves, southwest of Winnipeg, near Pilot Mound. His name was, appropriately, Frank Grain. How could I ever forget that? It was my first experience away from home among strangers. I was advised about work clothes, dress for travel and Sundays, and schedules on a CPR branch line. Near La Riviere I changed to a tiny mixed train (freight and passengers) to be met at Purves by Mr Grain with a wagon and team. Motor cars were rare in those days.

The Grain family had three young children so Mr Grain needed another person to help with the farm work. Andrews described the general routine at the farm:

> I was given a room upstairs and ate with the family in the big farm kitchen. My introduction to work began at 6 am the day after arrival. Briefly the routine was: out to the barn sharp at 6; feed and water the animals;

The Grain family.

clean out the stalls, moving the manure to a pile outside the far end of the barn; harness the work horses; milk one or two cows; then go back to the house for breakfast at 7. Before 8 the work horses were led out and hitched up for the day's work.

At noon it was lunch for man and beast. Horses were allowed only a small drink when they came in hot, but could fill up before going out again for the afternoon. At 5 pm we headed for the barn and again the horses were given only a small thirst quencher hence they were led into the barn to be unharnessed and fed. Cows were milked. Family supper was at 6. Evening chores included watering the stock, currying the work horses, bedding them down with clean straw, and filling up their mangers with hay plus a ration of oats. By 9 we went back to the house, to clean up, have a biscuit with tea or milk and off to bed. On Saturday afternoons and Sunday only chores were done. Insomnia was never a problem....

Frank taught me how to harness, hitch up and drive the team. He was patient and understanding. Previous jobs had taught me to pay attention, to remember, and to sort out problems....

A lifetime benefit of my SOS experience has been an understanding and love of horses....

I doubt that the Grains had previous experience with a teenage boy in the family. The novelty and fatigue were such that I did not even think of a bath for myself and must have begun to smell too! So on a Saturday, after supper, I was given the kitchen along with a tub near the stove for a quick bath. This became routine.

During the summer a tornado came through the farm:

One lovely July afternoon I was ploughing far out on the bald prairie. It was this work at its best, horses pulling well and just enough breeze for comfort. Going up wind the rich smell of horse sweat prevailed. Down-

Andrews as a Soldier of the Soil.

wind brought the aroma of freshly turned sod and perfume of distant hay meadows and wild flowers. Suddenly it became mysteriously calm, the horses began to fidget and I noticed on the horizon a strange dark cloud with serrated fangs piercing the blue sky above. It then circled all around, but mostly in the direction first seen. The horses showed symptoms of panic, so I unhitched and swung them around to face what I thought would be downwind. The dead calm continued but we could hear a crescendo of roaring wind.

Then the tornado struck:

Day turned to night. Horses, with their rumps windward, braced themselves with heads down. I crouched down in the same attitude, head between knees and hands over head. Breathing was laboured and flying gravel stung like buckshot. In moments which seemed eternity my thoughts sped back to pastoral ancestors, caught in prehistoric siroccos.

As suddenly as it started, the wind abated, the dust settled, daylight returned and we could breathe again. Like a bad dream the storm receded in the distance. The horses relaxed and ploughing resumed as if nothing had happened.

Although Andrews signed up for three months he stayed through the harvest, which began during the latter part of August.

[But] in mid September, while threshing was still in progress, I had to get back to high school, especially as I had missed the last three months of Grade 9. I had served as a Soldier of the Soil for 5½ months, nearly double what my contract required. Frank gave me honourable discharge, which validated my entry to Grade 10 in Calgary. The authorities gave me the bronze Soldier of the Soil badge, which sadly, is long since lost.

In 1987, while writing his reminiscences of his time as a Soldier of the Soil, Andrews and his friend, George Newell, went to visit the Grain farm.

William, one of the children at the farm when Gerry was there, still lived at the residence.

> We could not help arriving at 5 pm, an awkward time for unexpected
> visitors, recognized the old farm house at once, and the old barn which
> had been face-lifted with a new roof and coat of paint but the sur-
> roundings had changed. Lovely shade trees had matured and neat hedges
> bordered green lawns and flowers. Bill, the little 4 year old of 1918, now
> retired in his 70s, welcomed us and introduced his sweet wife, Margaret.
> Luckily I had with me the four photos taken in 1918, which were of great
> interest.... When I went upstairs to freshen up in the bathroom, I noticed
> along the corridor the very room in which I had slept so long ago.

According to his daughter, Mary, Andrews had fond memories of his time working on the Grain farm, and he was proud to have been one of the participants in this important wartime program.

While Andrews was a Soldier of the Soil, his father obtained a job as a commercial salesman for a pharmaceutical company. Since he travelled throughout western Canada, the family moved to Calgary, a more central location. When Andrews completed his service as a Soldier of the Soil, he returned to Winnipeg and met his father, who was there to finish the move to the new residence in Alberta. They drove across the prairie dirt roads. One evening they got lost in the Cypress Hills area and Andrews remembered sleeping wrapped up in the buffalo robe his father had in the car.

Andrews arrived in Calgary during the flu epidemic. His step-mother was the only person in the family to become ill, but she soon recovered. From Calgary Andrews saw the Rocky Mountains for the first time.

> On my way to school on cold winter mornings from the brow of the hill
> I was fascinated by the jagged horizon of the Rockies to the southwest
> across the Bow River valley.... Illuminated in rosy tints by the rising sun
> was a jumble of ghostly peaks, castellated cliffs and icefields. The myster-
> ies there and behind them were a challenge as they had been more than
> a century earlier for LaVerendrye, Mackenzie, Thompson, Peter Pond and
> ilk, in our history books. I did not know then that my professional life in
> British Columbia would be largely devoted to mapping and revealing the
> mysteries behind that panorama.

The Andrews family continued to struggle financially, so in the summer of 1919 Gerry searched for employment. He wanted to work away from home again, and particularly desired to see the land west of the Rockies. Later, he wrote about his work during the summer of 1919 in an article appropriately titled "Beyond those Rugged Mountains".

When Andrews spotted an ad for a waiter at the Railroad YMCA in Field he sent a telegram instead of a letter. "Even at that tender age I had learned that a telegram goes to the top of the pile for attention so I applied

Field, 1919.

by Night Letter. It worked. I went up to Field by CPR, the last Saturday in June – about a five-hour trip – through Cochrane, Morley, Banff, Lake Louise, over the Great Divide and down through the Spiral Tunnels to Field – exciting for a prairie lad!" Andrews described the place where he worked.

The "Y" at Field in 1919, was housed in Mount Stephen House, the old CPR hotel, built in the late 1880s. It was an enormous wooden ginger-bread structure, fronting on the station platform. Its coffee shop, open all hours, catered mainly to railway crews, and to day-coach passengers who could get a quick cheap snack there....

I reported to Mr Rice, the "Y" Secretary, and was given a small bed-room, one floor up overlooking the station. I was put on night shift in the coffee shop – 10 pm to 8 am, seven days a week, pay about $40 per month, all found. This was nearly twice what I got as a "soldier of the soil" in 1918. The seven-day week seemed hardly compatible with the "Christian" in YMCA. I soon got into the routine, if a bit clumsy at first....

A Chinese cook presided over the huge gloomy old kitchen behind. In quiet hours after midnight this area was in darkness except near the cook stove. At such times, to carry an overloaded tray of dirty dishes to the sink, I momentarily switched on the lights. This revealed hordes of cockroaches galloping across the concrete floor to cower in dark corners....

Andrews explored as much of the area around Field as possible.

I contrived to eat some breakfast before going off duty and supper after going on at night – but without much appetite. Often I made a bag lunch to eat outdoors. The quiet hour was about 4 am when I could hardly keep my eyes open. But by 8 am I was very wide awake. Instead of going to bed I had to get out in the glorious sun and scenery. There were easy hikes to beautiful and interesting places. I felt hemmed in by the four nearby mountains, Stephen, Dennis, Burgess and Field, which cradled the town. The problem, aggravated by the noise of trains below my window, was to get enough sleep....

One day a friendly pusher engineer asked me if I would like a ride with him up through the Spiral Tunnels. He said I should be near the track about ten o'clock, out of sight just beyond the station. I was there and as he passed, I hopped on the step and climbed up into the cab. The fireman, not much older than I, sat on a leather cushion on the left with his hand on a fuel control – pretty soft!.... In about half an hour we saw the beautiful Yoho Valley on our left and then entered the lower spiral tunnel. Lingering smoke from the front engines was suffocating and I fainted, partly from nervous tension. They revived me with a cold air jet and thought it a big joke. I survived the second tunnel.

During the summer, Andrews changed employment:

Toward the end of July I became disillusioned with the job at the "Y" – not enough sleep, no days off, smelly indoor work and poor appetite, but I did not complain. Someone must have recommended me to Joe LaBelle.... I had no contract with the "Y", so when LaBelle offered me a job as bullcook at the CPR tent camp at Takakkaw Falls up the Yoho Valley, I accepted and notified Mr Rice that I would quit as of the end of July....

The campsite, in a rough meadow commanded a fine view of Takakkaw Falls less than a mile away across the valley. There was good forage for horses and a small brook provided excellent water. There was a good cook tent, a large dining tent and about ten bedroom tents for overnight guests and like tents for the staff. A large tepee was used for evening campfires. The cook, Miss Pirie, was boss. She was very Scotch, mature, buxom, capable and short tempered. But she had a warm heart withal.... Meals served to guests in the dining tent were right up to top CPR standards of

Miss Pirie.
Right: Gerry Andrews, bullcook and
wrangler.

the day – spotless linen with the full array of dishes and cutlery. The two
women [the cook and waitress] enjoyed decorating the tables with wild
flowers. No liquor was served and everything was prepaid in Field so no
cashier was needed.

My duties as bullcook included cutting firewood and kindling, lighting
all fires, fetching water, washing dishes, peeling vegetables, burying non-
combustible garbage and keeping the premises tidy. I slept in my own tent
and ate at the cook tent. I had no direct contact with the guests....

My jobs required no supervision so I arranged my own timetable.
There was lots of fresh air and the environment was beautiful. In afternoons, after the lunch guests had gone and before the overnight people
arrived for supper, we had our interlude of spare time. Miss Pirie, on the
other end of a crosscut saw, often helped me cut down a dry tree for the
woodpile.

In early September, just before the camp closed, Thornton Andrews came
through Field on a business trip and spent a day with his son. "Brewsters'
Mr Currie kindly arranged a ride for him in one of the buggies coming
out for lunch. He enjoyed one of Miss Pirie's wonderful meals also on the
house and we had time to see some local sights before he had to go back.
He loved it." At the end of the season Andrews was given a saddle horse for
a few days with instructions to deliver it to the Brewsters in Field. He used
this opportunity to take a trip up the Yoho Valley, over Yoho Pass to Emerald
Lake, and then by road to Field, where he returned to Calgary and Grade 11
of high school. "I had now seen a bit of what lay just beyond these rugged

Fred Haggman.

mountains visible on the western skyline from Calgary. But I did not know then that the summer after next, 1921, Fate would give me one more wonderful summer based at Field, BC."

Gerry completed his junior matriculation in Calgary in 1920 and then went to work on a relative's ranch near Empress, Alberta. In the fall he joined his parents, who had moved to a cleared but undeveloped parcel of land in east Burnaby, a suburb of Vancouver. "My accommodation was a 9 x 12 foot tent on a lumber floor and sides, with an air-tight heater, bed and wash basin. I was quite happy and comfortable in it." Andrews enrolled in a general arts program at UBC for the 1920–21 school year.

After university finished in April 1921 he worked temporarily with Fred Haggman cutting firewood. Haggman, who was about 15 years older than Andrews, taught him how to use a cross-cut saw and double-bitted axe. Andrews wanted to return to Field for the summer so he wrote to Lyle Currie, the manager of Brewster's Guide Outfitters in Field. He got a job that included cleaning the stable, taking care of the horses and driving the wagon to take customers' baggage to and from the station.

> About mid July there was a temporary rush at Glacier House, west on the
> CPR mainline, between Golden and Revelstoke, where Brewsters also
> had the concession for transport. I was sent there by train and again was
> put on the baggage wagon between the station and the hotel.… Cleaning
> the small stable was also on my menu. I often did this in the evening and
> dumped the manure with a wheelbarrow into the Illecillewaet River from
> a bridge nearby.

After about ten days he returned to Field.

Andrews' father came to visit him again at Field. "Lyle Currie provided a free trip for us to Emerald Lake with a party of tourists. It was another happy interlude for Dad." Andrews spent some time at Wapta, east of Field,

Glacier House and the massive Illecillewaet Glacier in 1921.

as wrangler on day and overnight trips to Lake O'Hara, a popular destination near Lake Louise. "The highlight at Wapta was when Ray [Legace, one of Brewsters' guides] and I were assigned to take J. Murray Gibbon, chief publicity agent for the CPR from Montreal, and his party with all its paraphernalia for an overnight trip to Lake O'Hara." Andrews got to play in a movie scene they filmed. "One 'set' was for me to saunter down to the edge of the lake, gaze intently at beautiful Mount Lefroy rising steeply behind. After quenching my spiritual thirst, I was to stoop and gracefully dip a drink from the lake in the brim of my hat, smack my lips and resume my ecstatic gaze at Lefroy. This was my first and last ever role as movie actor."

Andrews and Legace rolled their blankets on the ground under a tarp while the guests slept in a nearby cabin.

> Just before daylight we were roused by a rumpus in the cabin, with shouts "Mr Guide! Mr Guide!" Ray grunted an oath and said, "Come on kid, there may be a bear in there." We pulled on pants and boots and rushed over. Inside the cabin all were up – with a porcupine for company. They must have left the door ajar for ventilation. The artist was down on hands and knees trying to shoo Mr Porky out the door with a lighted candle. He got it too close and ignited the quills. POOF – a burst of flame and a cloud of smoke – leaving the animal's back bare as raw beefsteak. The poor little fellow was quickly made insensible – forever.

When Andrews got laid off in August, Currie arranged for him to work on the CPR maintenance crew at Field. One of his jobs was to handle an ice cart:

> When a train pulled in, I tried to locate my ice cart near where the diner usually stopped. Then with nice clean ice chunks in a bucket I climbed up on top of the car with a light ladder and dumped them down a covered chute. Usually I made several ascents and had to be quick about it. One

The road to Emerald Lake and Yoho National Park.

day when I was up top, the train began to pull out. The ladder fell down and I wondered how I would survive going through the Spiral Tunnels in that precarious position. Fortunately a train-man noticed my predicament and gave three long pulls on the signal cord for the train to stop.

At the beginning of September Andrews returned to Burnaby where he enrolled at Vancouver Normal School for the 1921–22 school year. One of his science teachers in Calgary had advised Gerry to study forestry for it combined scientific knowledge and working outdoors. Gerry thought that teaching would provide sufficient income for him to someday return to university for the forestry program. There was a demand for teachers in the rural schools and it took only one year to qualify. He was particularly interested in a rural school because the salary was higher and he would have the chance to supervise his own school.

In the summer of 1922 Andrews worked at Fraser Mills lumber yard in Coquitlam while applying for a teaching position. His family returned to Manitoba in August, so he moved in with Fred and Kate Haggman, who became life-long friends and were like a second family to him. Gerry described Fred as "a marvellous mentor [who] holds a cherished place in my personal gallery of heroes."

At the end of July the Registrar of Teachers in Victoria sent a letter informing Andrews that his application to teach at Upper Big Bar School northwest of Clinton had been approved. He eagerly accepted the position and was ready to begin his adventures as a rural school teacher.

Rural School Teacher, Summer Horseback Trips

Upper Big Bar, 1922–23 School Year

In a letter dated July 31, 1922, Mrs Neas, the Secretary for the Upper Big Bar School Board, described the situation at her school:

> You may have the school if you can put up with our conditions. You would have to bring your blankets, say a pair. There is bedstead springs, stove, table, chair and a few dishes in a small house right at the school house. Our teacher batches here. The rent for the little house is 3.50 per month, 2 months paid in advance or 5.00 with wood furnished. It is in the school grounds.

On September 1 Andrews travelled on the CPR train to Ashcroft where he took a hotel room. The next day he continued to Clinton. "The motor stage was a huge Winton Six open car, seating eight or ten passengers with their hand baggage. It was a steep climb from Ashcroft about 7 miles [11 km] to Cache Creek through arid sage brush country." Andrews stayed overnight at Clinton where he met Harry Coldwell, who operated a small store and post office at Jesmond and had the mail contract to Big Bar.

> After paying for my room and meals at Clinton, I could not pay Coldwell for transport and overnight at Jesmond. He kindly said that I could pay after I got my September pay cheque. Our route from Clinton (elevation 3,200 feet [975 m]) was 10 miles [16 km] up a small creek, then another ten up Porcupine Creek to its headwaters, known locally as Dry Lake (elevation 5,000 feet [1500 m]). From there it was 10 miles to Jesmond (elevation 3,200 feet). The total of 30 miles [50 km] was a full day, with the horses having so much uphill pulling.

Andrews arrived at Upper Big Bar the following morning. He met the Cover family, who would have three children in the school, and Mrs Neas, who had five school children. Andrews went to Snider's store where he

Some of the hopefuls.

bought some groceries. "Snider had hauled my trunk from Clinton with one of his loads of groceries, for which I owed him $2. He also said that I could pay this from my first cheque. The new teacher's credit was evidently good in the community." Then Andrews went to set up his cabin.

> The water supply was across the road in Big Bar Creek which flowed rap-
> idly and noisily. I got a pail of water and lit a fire in the stove. There was a
> pile of firewood available for the school and my cabin. I had brought my
> double bit axe, which Fred Haggman had taught me to use with some
> skill. My first steps at "batching" were awkward. Kate Haggman had given
> me some simple recipes. Anticipation of starting my first ever school on
> the upcoming Tuesday banished any loneliness. I got some blankets from
> my trunk and made up the bed. I then rinsed out pots and dishes and
> cooked my first meal.

Andrews spent Labour Day preparing for the opening of school the next day:

> I inspected the school, opened the windows, found the readers for the
> pupils and inspected the toilets. The one for boys had no door, but faced
> the woods which provided adequate privacy. I then found the large Union
> Jack flag, somewhat the worse for wear, and practised hoisting it up the
> flagpole by a rope and pulley.
>
> There were 12 desks facing away from the road. Each had an inkwell.
> This was long before ballpoint pens came into use. The teacher's desk
> was a large table with a pull-out drawer containing the school register, a

A pioneer family.

chair and book ends. A cupboard contained readers for the various grades, chalk, ink, pencils and crayons. Wall maps included BC, the world, Europe, Canada, the British Isles and North America. They were rolled up and stacked in a corner. The one in use could be hung on the wall. An ample, smooth green-tone blackboard was fixed on the wall facing the pupils and behind the teacher's desk.

The next day Andrews started his teaching career:

I rose early on September 5th and, after my domestics, put on a tie and jacket which was proper dress for teachers then. I then opened and aired out the school, hoisted the Union Jack on the flagpole, and was ready to face my first ever scholars. They arrived early, no doubt interested in meeting their new teacher. I was not their first, except for the beginners. Sharp at 9 o'clock, I lined them up in order of size, girls first, marched them into the school, and seated them in the same general order.

At that time, first thing each day we recited the Lord's Prayer, standing at attention. Next, I listed their names in the school register. There were seven girls and only two boys. Two were beginners, three in Grade 3, three in Grade 4 and one in Grade 8.

Andrews quickly settled into the routine of teaching. "On weekends I got outdoors for fresh air and exercise, to offset confinement in school. Neighbours sometimes invited me for a meal. It is a special treat for a bachelor to have a meal made by a mother who enjoys cooking." Andrews became good friends with Clarence Cover, a friendship that lasted for many years. There were four rural schools in the Jesmond area: Jesmond, Big Bar, Big Bar Upper (Andrews' school) and Big Bar Mountain, but Andrews had very little contact with the other teachers that year.

There was no electricity at the school or in Andrews' cabin, so coal oil lamps or Coleman lanterns were the main source of lighting.

> In October the nights lengthened so that I used one or two lamps for homework and domestics in my cabin and could take one into school for any work there.... As the days shortened in October, the sun was hidden behind the high skyline south of my school.... During the gloomy season, on weekends I walked the three miles [5 km] up to Jesmond where I could see the sun, pick up my mail and walk back. It was good exercise and it sharpened my appetite....

> I had a real scare that fall. The walls of my cabin were double, separated by some 2 x 4 uprights, and the space between was filled with sawdust for insulation. I was told to dump stove ashes on the ground outside along the walls and later when snow came, to add it to the ashes in order to keep out the cold, and this I did. One night I had just gone to bed with the light out. Suddenly I smelled smoke. At the front corner of the shack I could see a glow. Fire! I jumped up quickly and found the pail had only about an inch of water in the bottom. I splashed that on the inside wall, dashed to the creek and filled the bucket. I ran back and dumped that water on the outside ashes which were smouldering. I got a couple more buckets full of water and splashed them inside and out and the situation was saved. No fire!

> Luckily I knew the path to the creek well enough to make these fast trips without injury. But after that, I always kept a full bucket of water in the cabin before "lights out" and was especially careful that any ashes used for banking had no live embers.

During that first school year Andrews also had an unfortunate encounter with pack rats:

> When I had arrived at the school, the large flag was already quite ragged. In due course a new one came by mail from Victoria. I carefully folded it and put it away for the night, but next morning I was dismayed to discover that pack rats had eaten into the folds. When it was unfolded it was full of holes. I hastily stitched the holes and well as I could and the flag was used for the rest of the year.

The school inspector did not visit Upper Big Bar until the spring. In his report he noted that the desks were in poor repair and that there was a lack of materials and facilities at the school, but he made favourable comments about the teacher: "Mr Andrews employs good teaching methods. He is working hard, and has met with a good measure of success this term. His work is very promising."

Summer 1923

"An important feature of teaching was two months summer vacation," wrote Andrews. "After ten months confinement in school it was imperative to get outdoors for some adventure. In my case this took the form of trips with pack horses on wilderness trails."

Andrews began planning for his first summer trip before the school year finished:

> In late spring of 1923, anticipating my upcoming trip by pack train down the west side of the Fraser River, I bought a fine little buckskin pony which I kept in [Clarence] Cover's fenced pasture. Appropriately his name was Jerry. Of mustang stock, he weighed less than 900 lb. [400 kg], was sure footed, and very tough. I then purchased a cheap British Army saddle by mail from a store in Vancouver, and obtained a bridle and halter locally. My little pony had been captured by old Billie Louie, chief of the Canoe Creek Indians. He had broken him to ride and trained him to sidestep at a swing gate. There were many of these swing gates and when the gate was open enough, he passed through and then sidestepped to close it. This saved the rider from dismounting.
>
> My pony was a deep buckskin colour with a dark mane and tail connected with dark fur along his spine from the withers to his tail. The effect was very attractive. He could jump fences like a deer and was hard to catch.
>
> Cover helped me attach a short length of chain to one of his front feet with a leather anklet. This prevented him from jumping and made him easier to catch and did not interfere with his grazing. He likely realized that when I rode him, he got rid of the chain nuisance.

In June, shortly before the end of the school year, Mr Neas was convicted of stealing cattle. The judge gave him a suspended sentence provided he leave the area, so the Neas family returned to Montana taking over half of Upper Big Bar's school population. Andrews thought that the school would close and that he would need to seek another teaching position. He did not anticipate that he would return again in two years.

After school finished Andrews helped Clarence Cover with haying until July 23 when he left on his pack trip. Andrews kept a diary for most of his trip, later adding notes for the last few days. On the first day he travelled toward the Fraser River, en route bidding farewell to some of the friends he had made during the past year. One of them gave him a young collie named Pinto for company. In late afternoon Andrews crossed the Fraser on the High Bar ferry and proceeded south down the west side of the river. "Camped at spring 4 miles [6 km] below Watson Bar Creek. Poor feed, water alkali, and little of it." The next day, "after no sleep, got up at 3 am. Got away by 6. After five miles reached some Indian cabins at good spring." By evening, after about 40 kilometres of travelling, he had reached BC Marsh's place. "He invited me to stay overnight, gave me supper and let my horses feed in his alfalfa field. Slept in a straw sack. Got fleas!" On July 25 Andrews visited Charles Shaw, the teacher at Pavilion. In his notes he says that "to get across the Fraser River to see Shaw ... I used the 'cable ferry', then used for single passenger and/or mail. This was scary but very exciting!" About 5 pm Andrews and Marsh left for Lillooet, camped for the night at a springs, and reached Lillooet the next day.

From Lillooet Andrews travelled alone, heading west "along north shore of Seton Lake on PGE [Pacific Great Eastern Railway] grade. It is a beautiful lake, high mountains all around with steep side hills. Water is a 'mountain' green, clear but not cold." In the late afternoon, when he started looking for a camp site near Seton Portage "old Bill Duguid, working in his orchard saw me and invited me to put my horses in his barn, feed them hay, and stay the night with him and his young son, Jack. Talked with Duguid till midnight. A most interesting character." Andrews was impressed with his host's establishment:

> Duguid's place [is] well under cultivation, hay, garden, pasture and orchard irrigated by sprinklers. His house is well furnished with things of his own making. Most interesting is his hydro-electric power system. He has a 3 inch feed pipe from a spring 70 feet up, free running, winter and summer. In his shop, a Pelton Wheel which he made turns a small generator with enough power to operate a full load: lights in all buildings, cream separator, churn, washing machine, grindstone, etc. It has run for several years without a breakdown.

The next day Andrews reached D'Arcy at the west end of Anderson Lake after a long day of travelling that included crossing several trestle bridges:

> The horses and I were very tired. At any time, their feet could have slipped down between the ties and broken.... The final hurdle was a tunnel into which we had no alternative but to go. When we emerged, it was completely dark. Approaching us was a hand-powered rail cart with a light. It was the local section foreman inspecting the line for an eastbound train due soon. He expressed eloquent surprise that we had come all the way on the railroad. By then, the horses and I were dead beat, so I camped at the lake's west end, too tired to eat but we quenched our thirst from the lake.

From D'Arcy Andrews followed wagon roads to Pemberton where he found the area in flood. "I could not see the road, submerged in the milky water, but on each side the tops of the fence posts were visible. By keeping to the centre between them, we followed the road, the water coming to the horses' bellies. I carried my little dog, Pinto, up in the saddle with me."

Along Seton Lake.

At Pemberton he changed his plans:

By good luck, there was a prospector at Pemberton waiting for the flood
to subside so that he could get up into the hills. He offered to buy my
pack outfit for a bit more than it had cost me, and for cash. I accepted and
decided to take the next westbound PGE train to Squamish, then by the
old Union Steamship to Vancouver, and go on to New Westminster, where
my Haggman friends lived.

At the station when I bought my train ticket, the agent said that I must
also pay for my little dog to ride as baggage, a problem I had not anticipated. I couldn't possibly have her in the city. The agent seemed friendly,
as did his wife, who also appeared. When they offered to take Pinto I was
sure she would have a good home. My problem was solved.

My arrival in Vancouver by the PGE to Squamish and by steamship to
Vancouver was at night, too late to go on to New Westminster. Being still
in bush clothes, I decided to get a cheap room. An illuminated sign said
"ROOMS". The office was up a flight of stairs, where I found the "Ma-
dame". She wanted $3 "pay in advance", which was all right. Then she
said that for another $5 she would get me a nice young bed partner! I was
very tired, so it was easy to decline the "extra attraction". A colourful end
to an exciting journey to be sure! In my long life, I have certainly slept
in some queer places both before and since that incident, but have always
kept it secret that I once spent a night in a brothel!

In concluding his account of the trip Andrews included a final note about his dog, Pinto:

> As a graduate forester in 1931, I was temporarily doing forest surveys near Clinton and had occasion to go to the PGE station to get an express parcel from Victoria. The agent and I looked at each other, sure that we had met before. I then realized that he was the man who had adopted Pinto at Pemberton! He said he still had her and then whistled and she appeared.... My intuition at Pemberton, eight years before, that Pinto would have a good home had happily proven correct.

Kelly Lake, 1923–24 School Year

After returning to Vancouver, Andrews travelled to Victoria to meet the registrar of teachers where he expressed his preference for a remote frontier school. The registrar gave him two choices in the Peace River region. Andrews selected the teaching assignment at Kelly Lake, a Metis community near the BC-Alberta border that was starting its first school. He was still a teenager, but his experiences during the past year had given him confidence to try a new adventure in a remote part of the province.

He sent his application to the Kelly Lake trustees but didn't receive a reply until after Labour Day, when he was offered the position:

> Leighmore, Alberta – Sept 8th 1923
>
> Dear Sir
>
> I wired the Teachers Bureau there accepting you to teach Kelly Lake School if you wished to take it. I wanted to let you know what kind of a place and school this is.
>
> In the first place this school will be all Breeds, starting to school for the first time and not many can talk English. The people around here are mostly all Breeds and trappers. It is about 60 miles from Grande Prairie Railroad station. You would have to batch. I stay there most of the winter buying fur so if you thought you would care to live in a place of this kind it would be all right to come. I could meet you at Grande Prairie if you write me when you would be there or wire to Beaver Lodge.
>
> I had a friend here that wanted to get this school but he didn't have a teachers certificate for BC and couldn't get a permit to teach it.
>
> Jim Young

The school was to be located in part of the trading store Young operated at Kelly Lake.

On September 11, the Education Department in Victoria sent Andrews a confirmation of his appointment to teach at Kelly Lake School. To his dismay he learned that school wouldn't begin until October 1, a loss of a

Jim Young.

month's pay. In Vancouver, Andrews did some shopping for items he anticipated that he would need to start a new school. On September 18 he left
by CNR train for Edmonton, arriving there in a snowstorm. Then he took
another train to Grande Prairie. After staying overnight he took a horse
stage 50 kilometres to Beaverlodge. Jim Young met him there with a team
and wagon. It took two days to reach Kelly Lake, stopping overnight at
Leighmore, Alberta, Andrews' nearest post office for the next two years, and
he wondered if he might be "the only teacher in British Columbia whose
pay cheques were sent to an Alberta post office." Andrews commented on
an obvious landmark near the end of their trip. "About a mile from Kelly
Lake we crossed a conspicuous north-south cutline slashed through the
bush. This was the Alberta-BC boundary on the 120th Meridian which
R. W. Cautley, DLS [Dominion Land Surveyor], had surveyed and marked
just the year before."

Jim Young, who had purchased the trading store at Kelly Lake the previous year, wanted a school to be established at Kelly Lake, and was Secretary
of the School Board. His trade store was a large log building about 6 x 12
metres with a sod roof. A room in the east half of the building that had previously been used as a games area would become the new school. Andrews
and Young spent the next few days converting this part of the building into
a school. They used shiplap to build a partition wall. Andrews hung a blackboard cloth that he brought with him on the new wall.

> As yet there were no desks for the school. Jim had made a deal with
> the trustees at Tupper to get their old homemade units which had been
> replaced by factory made furniture. We had time for a quick trip to get
> these before the weekend.... By Saturday the new school in Jim Young's
> store was as ready as could be for operation; meanwhile the people had

Kelly Lake School.

returned from their berry picking trek. By good luck a Swede trapper
fluent in Cree (Joker Sanderson) came to the store. He cheerfully made
the rounds of Kelly Lake families to tell them that school would open
next Monday morning, 1st October, at 9 o'clock.

Young hauled goods for his store from the railroad at Grande Prairie and
departed on Sunday for a week-long trip, leaving the new teacher to start
the first week by himself.

Andrews described the beginning of the first year of school at Kelly Lake:

Up early completing my domestics, a bit after seven, I looked out the
window to see little brown faces furtively peering at the premises from
cover behind surrounding trees. These were my pupils arriving already.
Few families had or needed clocks.... If I waited to open school at the
statutory 9 am, the kids might lose heart and vanish – God knows where.
So, before 8 o'clock I stepped outside, properly dressed (tie and jacket),
hoisted the large Union Jack to the top of the flagpole and vigorously
rang the old brass hand bell which Jim had found somewhere. The
children, one by one, cautiously emerged from cover into the clearing.
Smiling, I beckoned them to approach, arranged them in line, girls first,
smallest in front, ushered them inside and assigned them to their seats in
the same general order.

There were about a dozen, nine boys and three girls. Ages ranged from
about 6 to the oldest, Henry Belcourt, a handsome lad of 14 and taller

Recess at Kelly Lake School.

Urban Gladu, a trapper and the father of Colin, one of Andrews' students.

Andrews' main social event was dancing:

> Jim and I were invited to occasional local dances. Everyone turned out
> in their best finery. In addition to the ubiquitous square dances, they had
> some native routines, which seemed like adaptions from old country reels.
> One was the "wapoosaywin" or "rabbit dance". There was ample local tal-
> ent with the violin for lively music like the "Red River jig".

Trapping provided the main income for the people at Kelly Lake:

> Winter snow did not begin to accumulate until early December that year.
> Before this, clear cold weather, prior to the snow, had allowed the frost to
> penetrate deeply and harden the ground. The late snow delayed the men's
> departure for their trap lines, so their return for the holiday break was also
> delayed. Our anticipated school concert was, therefore, postponed until
> after the New Year.

Immediately after school holidays started at noon on December 21,
Young and Andrews left with a sleigh load of fur for Pouce Coupe, stopping overnight at Peavine Lake. At Pouce Coupe Jim sold his furs and purchased some cases of hard liquor at George Hart's store. Then the two men
returned to Peavine Lake. When they got back to Kelly Lake the following
day, the two men stopped for lunch before heading to Leighmore, Alberta.
On Christmas Eve they drove to Grande Prairie:

> It was a long haul that day to Grande Prairie via Beaverlodge. Arriving
> after dusk, Jim turned into a back lane, paralleling the one and only main

street of town. About halfway along the lane, he opened a gate in a high board fence, to enter the back yard of Bud Lay's poker emporium. In this seclusion, we transferred Jim's cargo from Pouce Coupe into the rear of Bud's premises. He was obviously gratified that his holiday inventory would be in good shape, thanks to Jim. We were more than welcome to stay there overnight, but did linger next morning, Christmas Day. After a long and bitterly cold trip, we got to Leighmore in time for Christmas dinner in the evening with the Beadles [friends of Young].

The school's Christmas concert was held early in January:

Before the fathers returned to their trap lines, we staged the long-anticipated Christmas concert. We had a day or two to freshen up the repertoire, which had been practically ready before the holiday. These were simple items appropriate for each age group, to show off their scholastic accomplishments: short recitations, a group song or two and some simple dramatics. The schoolroom was packed, with barely enough room for the performers. Furniture backed along the walls provided seats for the elderly and squatting on the floor was no hardship for the rest. Winter wraps and one or two somnolent infants were stowed in Jim's store area.

My father ... had just sent me an ingenious little portable gramophone with some records, including one or two Christmas numbers. He suggested I might stage a surprise with it. Our concert was the opportunity. I hid it in a corner behind a large wall map, all set to go. At the end of the kids' program and just before refreshments, I sneaked behind the map and triggered the machine. Suddenly, the hymn "Joy to the World ... the Lord is come!" boomed out in rousing volume, almost lifting the sod roof. Eyes opened like saucers and jaws dropped down on chests. It was a smashing finale to our program. More gramophone selections were played during refreshments. Candy, biscuits, etc. from Jim's store were dished out with lots of hot sweet tea. They had brought their own mugs. The first ever Kelly Lake school concert was a signal success.

Andrews used local knowledge to make a robe that winter:

With rabbits so abundant that winter, someone suggested I make a rabbit robe. A local lady showed me how.... My robe used between 150 and 200 skins. Finished, it was like a furry blanket, about two inches thick. A finger could be poked through the loops anywhere. The same lady ... lined my robe with flannel and made the outside cover of canvas, with flaps, and moose-hide tie-strings. It folded into a wonderful sleeping bag. It was almost like a furnace, and I used it outdoors on the coldest nights. It weighed about 20 pounds, too heavy and bulky for a backpack, but OK for a sleigh. In summer it was too hot as a bag, but could serve as a mattress for two people. I paid my Kelly Lake school boys 5 cents per skin, processed into thongs. It was a bonanza for them.

G.H. Gower, the school inspector for the Peace River district, visited Kelly Lake toward the end of February:

> He arrived at Kelly Lake late one afternoon quite unexpectedly, having come in a light cutter with a driver via Swan Lake and Brainard's. Bed and board for them were no problem. Jim was away, so his cot was vacant. Mr Gower spent most of the next day in the school.... In the morning he had me go through my usual routines with younger and older age groups both separately and together. After lunch he took over with his own questions and routines.... He strongly urged me to return a second school year, so our good start would not be in vain.

In his report, Inspector Gower commented on conditions at the school and Andrews' teaching:

> The school in this district occupies a portion of a large log building. The classroom is neither well lighted nor ventilated, and there is but a small area of slated canvas blackboard. There are no regular school closets. Home-made desks are in use. Water is available. The regular set of maps, supplied by the Department, is in the school....
>
> When the school opened last October, for the first time, hardly a child had any knowledge of the English language. Under Mr Andrews the children are acquiring a knowledge of the language and are progressing favourably along other lines. Mr Andrews is well adapted to work of this type. He has an abundance of patience and shows skill in the teaching and management of non-English speaking pupils. His work in this district is very commendable.

By the end of the school year, Andrews wrote: "In general all scholars were in good shape to resume studies after summer holidays, when the curriculum would broaden to include geography, history and more advanced English."

He had one more thing to take care of before he could begin the summer pack trip he had planned. In late spring, Inspector Gower had written to him advising that he apply to the provincial government to write the senior matriculation exams at the beginning of summer. Andrews applied right away but had not received a reply by the time school closed on June 19. He went to Leighmore with Young on the following Sunday, where he received a telegram from the registrar in Victoria telling him to report to Rolla at 9 am on Monday, June 23. "This was tomorrow and more than 80 miles [130 km] distant by trail."

Andrews immediately returned to Kelly Lake on horseback, where after a brief rest he started his journey:

> [I] took the old surveyors' trail along the Alberta-BC boundary to Swan Lake. This was more direct than via Peavine Lake, and I had hiked part of it before.

It was practically the longest day of the year and our direction was due north. We sleuthed our way along the faint trail, detouring for the odd muskeg and windfall, dropping into and climbing out of occasional ravines. I could verify our whereabouts by the prominent boundary monuments at intervals of about a mile. It was nearly 11 pm when the sun finally dipped below the horizon, just west of north. Then its twilight glow moved slowly to the right across our path ahead. The horse seemed to share my sense of urgency. He behaved willingly and well. As the twilight darkened we stumbled along with too many blunders. Horse and rider felt the magic of the semi-darkness, when bush, trees and boulders assumed weird and occult forms. Only our own sounds broke the mysterious silence. Gently the sun reappeared ahead to right, promising another beautiful day.

About 3 am from a commanding hill we could see Swan Lake about three miles [5 km] ahead, and 500 feet [150 m] below, reflecting the full glory of the dawn. I soon reached Ma Flynn's where all was dead quiet. I stabled and attended the horse, removing the saddle to air his back, then crept into the house to find a spot to lie down. Ma roused to ask "Who's that?" and told me to find a bed until breakfast time – which came all too soon. I then saddled up again.

We stopped briefly at Pouce Coupe to refresh ourselves, and arrived at Rolla early afternoon. By this time both horse and rider were getting a bit groggy. However, I reported immediately to the presiding teacher at the school and explained the circumstances of my late arrival.

I had already missed the first exam. But as I was the only candidate for Senior Matric, she said I could write it next morning and the other exams immediately following, instead of taking the whole of the week as per schedule – provided I would go right back to the hotel and to bed. May God bless her soul, sensible and kind.

He passed all his exams.

Andrews' first year at Kelly Lake had been successful. He had developed a close friendship with Jim Young and a good rapport with his students and the adults in the community. He enjoyed the adventure of living in a remote community in northern BC. In summarizing the year, he wrote, "It had been most worthwhile in all respects and I was quite ready to follow Mr Gower's advice and return there for another year."

Summer 1924

"Long months of confinement in store and school at Kelly Lake having sharpened our appetites for some action and adventure, Jim Young and I planned a packhorse trip over Pine Pass." Young had heard about good live-stock in the country west of Prince George and he hoped to sell his horses there after the trip. His friend, Fred Barber from Beaverlodge, came with them and brought along his two dogs. "We decided on six horses – three to ride, two for packs and one spare – and to assemble and outfit at Kelly Lake in late June." Young provided the horses, Barber brought his fishing tackle and Andrews took a camera and rifle. They took no tent but planned to use a large canvas pack cover for shelter. The food was mainly dry staples. The men spent a couple of days at Kelly Lake assembling and organizing the outfit before departing for the five-week trip on July 1.

Andrews kept a diary of this trip and later added notes. The men travelled north about 40 kilometres to Swan Lake where they stopped for a day and made panniers for the coal oil crates that they were using for packs on the horses. Andrews had learned how to make these when he was at Big Bar. On July 3 the group continued north to Pouce Coupe where they did some shopping. "After considerable beer, the party turned in at Fynn's Feed Stable." In Andrews' notes he described the group's departure from Pouce Coupe the next morning as they began heading west:

> The sun was high in the sky when we finally got away from Pouce
> Coupe. In packing there is always some initial confusion until the best
> routine becomes habitual. An interested audience was in attendance,
> mostly trying to be helpful. Our lack of experience must have been very
> obvious. Advice from spectators flowed freely, including various ways to
> "throw the diamond hitch". When we finally got away our conception of
> it was a fearful tangle of cinches, lash ropes, straps and tarpaulins, with odd
> bits of equine anatomy sticking out from a nucleus of confusion. Good-
> natured farewells were exchanged, and there was a twinkle of humour in
> more than one pair of eyes left behind. There was frank admiration, too,
> for our tackling what was reputed to be a tough trip, also one or two sar-
> castic predictions that we would be back in a couple of weeks, out of grub
> and beaten! We were glad when we got on our way, at last, each leading a
> packhorse, and heading into the challenge of the unknown ahead.

The men travelled through Dawson Creek, their route to the Pine Pass closely following present-day Highway 97. On July 6 they crossed the rivers at East Pine just after lunch, with some assistance:

> Dropping down the long hill, we got to East Pine about 11 am. Here
> the Murray River from the south joins the main Pine flowing from the
> west. A friendly old-timer, Jim Daniels, advised us to cross the two rivers

separately, above the forks, to reach the main trail up the Pine on its left (north) bank.... A squatter at East Pine named L.C. Palmer helped us make the crossing with his dugout canoe, for $3, which saved making a raft. The horses swam both rivers. Palmer gave us good advice for a camp on the far side and showed us his simple "diamond hitch" which we used thereafter to good effect.

On July 7 they camped at Centurion Creek about a kilometre south of the present town of Chetwynd. Two days later they reached Philip Esswein's ranch. "A trail from Hudson Hope joins the Pine Pass trail here.... Esswein and Neaves [a local trapper] say another 25 miles [40 km] of good going, then 35 miles [55 km] of bad trail to summit.... They say it is a week's pull from Esswein's to the summit." By July 11 the men reached the mountains.

In the morning, no horses. Jim went back several miles for them. Got started at noon. Trail bad so crossed river and advanced on a sand bar, then recrossed to a small flat. Found hidden slough mentioned by Neaves, about 50 yards to right of trail, just after the old trail crossed a little bridge. Made camp on the trail. A good place but not much room. Nearly half the rum left. Barber says it must be finished before the summit. Mountains getting higher and closing in.... First devil's club seen. Jim found a large horse bell in the slough. It will be a boon.

The next day they reached Neaves's cabin at Callazon Forks. Then the weather turned rainy and the trail became more difficult to follow, which slowed their progress toward Pine Pass. On July 14 they passed another of Neaves's trapping cabins.

Lost trail badly after leaving cabin, but located it again after hours of wet scrambling in heavy shower. Made camp at 6 pm; poor feed, but some. Seem to be on the right trail, so are in hopes of a good day tomorrow. Weather looks bad. All got a good soaking today, but Jim could not wait for the rain, so fell into the creek at noon....

Rain – rain – rain – all night, all morning, all afternoon. Sat around a big fire all morning, but after lunch, packed up in the rain and pulled out. Pushed along for about 5 miles [8 km] till we came to a trapper's cabin. Stopped here for the night. Weather looks like clearing.

They remained in the cabin for the next day:

Still raining today, but had a good night in the cabin. Spent the morning scouting for the trail. After lunch Jim and Gerry went ahead to clear out the trail for a mile or so. It is very indefinite and bad. Weather looks like changing. The old sun came out for a few minutes at 6 pm. Creek is high and murky due to heavy rains. The mountains take on a more severe aspect here.... We are cheerful and hope to make the summit tomorrow. It cannot be far away now.... Made good use of the trapper's cabin to dry our wet clothing. Sugar is practically gone, also bacon and ham. Lots of flour, beans and rice, should be enough to see us to McLeod. Andrews later recalled:

I think it was at this camp that my companions used up the last of the rum. They insisted I participate in this final toast to our success and survival. They decanted my modest portion into a mug, offering hot water for mixer, which I declined, affirming that if I had to drink rum, I would have it straight. Choking violently, I wasted most of the precious nectar coughing and sneezing, to the dismay of my generous hosts.

Jim Young frying fish at Pine Pass.

The men reached Pine Pass in the mid afternoon on July 17 by following a faint trail along the east mountainside. In the pass they found Azouzetta Lake and a trapper's cabin at the south end. They rested for a day at the cabin and got cleaned up. Barber caught 24 small rainbow trout. In Andrews' notes he recalled their stop at Azouzetta Lake:

Arrival at Summit Lake (Azouzetta) in the pass was the high point of our trip, both topographically and spiritually. We were lucky to find feed and a congenial camp.... [Axel] Rosene's cabin was beautifully built of peeled logs and quite new. The sod roof was a mass of floral colour, and thanks to our arrival, the pansies had not been "born to blush unseen".... His name and origin (Prince George) were inscribed in bold letters above the door. It cheered us to think that Mr Rosene had travelled the route we were now to take. It was also reassuring to know that we would now be following downstream drainage.

The three men left Pine Pass early in the morning of July 19, descending to the Misinchinka River. In his notes Andrews wrote: "On the trail down from the lake, we noticed with interest that the more recent blazes were all 12 feet or more [about 4 m] above ground on the trees. This implied they had been made in winter on top of at least eight feet [2.5 m] of snow! It confirmed that recent travel had been on snowshoes with no need for a summer horse trail." They found another small cabin of Rosene's and stopped there for the night.

The men could not find the trail, even after searching for it all the next day, so they remained at the cabin for another night:

At supper we held council of war, debating question. Finally decided to
go down in the morning where we turned round yesterday, try crossing
the river for a trail on the other side. This was the last straw. Grub enough
to reach Cutbank should we turn around, and probably enough to reach
McLeod if we have half-decent luck. Figure it is not more than 40 miles
[65 km] to McLeod. Our grub should last 7 days more. If we find a good
trail we are OK.

Andrews later commented:

Here we were at the moral "summit" of our trip – the point of no return.
All were loath to admit defeat and face the humility of returning beaten
to our Pouce Coupe friends.... Jim Young's determination and guts tipped
the scales for going ahead, no matter what. With typical generosity, he said
we could butcher one of his horses if we got to such extremity.

On July 21 they proceeded downstream. "Got down to the crossing and
had lunch there at 11:30.... Found a cabin and a trail on the other side (the
right or west bank). Followed down river in pm. Trail very indefinite, over-
grown with alder and willow.... Camped on a sand bar for the night. Jim
found the trail on the right bank." Although the trail was difficult to follow
the men knew they were travelling in the right direction. The next day they
found a large cabin. "Had the usual trouble picking up the trail beyond the
cabin, but finally came to some good grass on a willow flat. Camped here.
River is quite wide and meanders considerably." Andrews noted: "We noticed there a large 45-gallon drum and other evidence that it was a trapper's
headquarters at the head of river navigation from McLeod Lake. This implied that horses were not used for transport, certainly in recent years, and
explained our trail troubles. Winter travel on snowshoes over deep snow is
not impeded by soft ground, water or underbrush."

Two more days of travel brought the group to the Parsnip River where
they found an old cabin. They camped on the east bank of the river, built
a raft and prepared to cross the river in the morning. Andrews recalled this
campsite:

We had now entered the huge valley of the Rocky Mountain Trench....
Our camp was pleasant, at the mouth of Colbourne Creek and facing
west across the big river. We added our names with date to many others
inscribed on the unattended "guest book" on the walls of the old cabin.
It was cheering to find so much evidence of humanity after all the lonely
days since Esswein's. This location had the indications of a crossing point
and tallied with my map. At low water it was likely a ford, but now the
river was running too high from glaciers melting in the midsummer heat.
We were on a historic river route of explorers, prospectors, surveyors,
trappers et al. That pleasant evening, looking across the Parsnip, reflecting
the glory of the evening sky, I was thrilled to think that here, just in front

of us, Alexander Mackenzie and his voyageurs had passed in their ascent and descent of this river 131 years before.

The next morning the men crossed the Parsnip:

There was plenty of dry driftwood along the river edge for our raft, to carry two men and all the gear in one trip. We had lots of rope to bind the logs together. Looks did not matter as it would be a one-time conveyance. With the strong current we had to allow for both raft and horses drifting downstream some distance. Exposed gravel banks on both sides offered ample room for departure and landing. I was elected to ride the lead horse across, being the only one to admit he could swim. Actually, I felt better about swimming on a horse than riding on the rickety awkward raft. When all was ready, I mounted the grey and, leading Elsie we plunged in. Just as my horse began to swim I realized the saddle cinch had broken – not so good! Luckily we got turned round and regained the shore. The cinch was soon repaired, but I also realized the grey was not a good swimmer. Too lanky and raw-boned for buoyancy, he would be at a disadvantage with me on his back. We transferred the saddle to Elsie, shaped like a canal barge. This time, leading the grey, we started in again, Jim and Barber chasing the other horses to follow, loose and without loads. All went well. Elsie swam high and I got wet only to the waist. It was a relief to feel her feet strike bottom on the far side, about a quarter mile downstream. I was then able to catch and tie the other horses as they emerged. After that, it was my turn to be spectator as the others crossed on the raft, working frantically with clumsy improvised paddles. The crossing was a success – nothing lost and nothing wet that mattered. It could have been otherwise.

From the Parsnip the men got about halfway to Fort McLeod before stopping for the night:

Our rations were practically exhausted. While Jim and I set up camp, Barber wandered off with the .22 and soon returned with a porcupine, skinned and nicely dressed, ready for the pot. I suspect it was a grandfather. With no grease for frying we boiled it. Broiling might have been have been better.... All we had for breakfast, after this repast, was a handful of flour for a tiny bannock and a pinch of tea for a weak brew. Good thing we had filled up on the blueberries earlier.

On July 26 the three men arrived at Fort McLeod:

When our outfit broke out of the timber on the lake shore, near the outflow down Pack River, a flotilla of native canoes crossed at once from the little settlement opposite, to greet us with obvious surprise and curiosity. "Where you come from?" they shouted. We countered with, "Where can we cross with the horses to get to the store?" They indicated an easy ford across the Pack River nearby. We soon drew up beside the HBC depot and dismounted amid great local clamour, especially from the children.

I asked one friendly old man, who spoke a little Cree, why the kids were making such a fuss. He replied, "Him never see him horse, him think you ride funny moose!"

The HBC manager, Justin E. McIntyre, proved a great character. I got to know him well in later years. We entered the store at once, having nothing whatever with which to make our lunch. Mac looked out the window at our outfit and repeated the Indians' question. When we mentioned the Pine Pass trail from Pouce Coupe he said, "You've had a tough trip!" – and disappeared through a door behind the counter. He returned with four glasses and a bottle of rum, poured a hefty slug in each glass and said, "Your health, gentlemen!" He explained he had worked on the PGE railway extension survey up the Misinchinka River a few years earlier. He insisted there was no feasible route for horses directly to Prince George, but we should have no trouble on the old trail to Fort St James, which had been a main thoroughfare for horses in earlier days. I met McIntyre again in 1948 and often thereafter. By then he had set up his own store on the lake's east side, accessible from Hart Highway. He was a great favourite with all survey crews using this route to and from the North.

Andrews and Barber attended mass on Sunday morning, July 27, before leaving Fort McLeod:

There was no priest. A native elder led the service, prayers, hymns, etc. in their native tongue, Sekani. It was evidently a devout band, reflecting the enduring influence of Father Morice, some 20 years after his retirement from Fort St James Mission. The congregation squatted on the floor or stood at the back, women and infants on the left, men and boys on the right. The service lasted about half an hour.

The trip from Fort McLeod to Fort St James was easy and the men made good time. The trail was well-cleared and blazed and there was an abundance of feed for the horses. The group passed Carp Lake and Carrier Lake. "Somewhere along this part of the trail we came upon a tiny lonely grave in a peaceful little glade. It was marked by a crude wooden cross, mute for name and date but eloquent of grief and affection." They reached Fort St James in three days, then two more days of travel brought them to Vanderhoof on August 2. "Found fair grazing at Vanderhoof. No demand for horses." They spent almost two days there before leaving for Prince George on August 4. Jim had traded his outfit for a 1921 Model T Ford touring car. Barber had already gone ahead by train.

When they reached Prince George, the three men decided to drive to Vancouver. They proceeded south to Cache Creek, then east to Kamloops and down through the Okanagan, crossing into the USA.

At Oroville, the US Customs looked with disdain at our uncouth outfit across the road, and contemptuously waved us through.... We stopped two or three days at Tonasket to replace a broken crankshaft and stayed at the quaint old hotel there....

It was after mid August when at last, we arrived in Vancouver. There we settled mutual accounts, and each went his own way, I to the Haggmans' in New Westminster for a short holiday, before heading north again to rejoin Jim Young at Kelly Lake and to get organized for another school year there.

Many years later Andrews recalled that summer adventure:

For me it had been an ideal vacation, even if a bit more expensive in time and dollars than I had planned, due partly to my share in repairs to the old Ford car at Williams Lake and Tonasket. I had enjoyed the companionship of two very fine senior friends in an exciting and arduous adventure. There had never been a moment of discord among us. I had been the target of some good-natured teasing, especially from Barber, but I know now that this was a genuine expression of regard. For me, both geographically and historically the trip had opened wide and wonderful vistas.

Kelly Lake, 1924–25 School Year

Andrews returned to Kelly Lake about a week before school started. This year the school would be held in William Calliou's vacant cabin. The cabin was in good shape, but Young and Andrews needed to do some work to convert it into a school. After they completed the work in the school, the two men changed the old school at Young's cabin into more spacious living quarters for themselves.

All of Andrews' 12 students from the previous year returned, along with 6 more. Just like last year, none of the new students could speak English. But the beginners had the advantage of their teacher's experience in giving language lessons and their schoolmates' knowledge of English. During the early fall Andrews built some desks and benches to accommodate the additional students. He was much more confident about teaching at Kelly Lake this year because he knew most of the students and was familiar with life in the community.

Andrews also continued night classes for a few adults. He described these in a letter to his sister, Nora, in October: "I have just dismissed my 'evening class' consisting of four lusty dusky fellows of 18, 19, 23 and 28, the last two married and raising families. It is about 10:30 pm. You see we have a night class once a week for these four ambitious fellows who want to learn to read, write and figure. Perhaps I can get them started anyway."

Unlike the previous year, permanent snow came early in the fall of 1924. "Winter imposed his stern discipline on our world. In real cold snaps, with temperatures down to −50°F or lower, invariably the air would be clear and deathly still, with smoke rising straight up from chimneys in a dense thin vertical column." Andrews also had to get the school ready for the students. "During winter, before breakfast, it was my habit to go across to the school and fire up the big heater so that as the kids arrived, they could enter and get warm. By 9 o'clock the building would be sufficiently comfortable to start lessons. We lost only a couple of days from extreme cold."

In a letter to Nora, Andrews described how he spent Christmas:

The day before Christmas I hiked to Peavine Lake – 15 miles [24 km] north, where Alec Anderson lives. Xmas Eve in a trapper's cabin, listening to his stories of youthful pranks in good auld Scotland – between these and smokes, my mind travelled far in time and space.... I was very happy!

Christmas morning, before daylight, Alec the trapper and I started off to Swan Lake, 9 miles [14 km] farther north. It was a good walk, and we got to Frank Ward's just in time for our "trappers" Xmas Dinner – a goose, nuts and candy – and bushels of jollity. Our nine-mile appetite didn't leave much of the goose. Alec, Ward and Miller, all bachelors and trappers, are fine men. In the afternoon we took turns running down to the lake to catch fish through the ice. I got three dandies. About 6 o'clock all four of us tramped over to "Bone" Taylor's. His good wife had invited us for the evening meal. Here – more goose, turkey, plum pudding, Xmas cake – such a feed! They have five girls (the youngest less than a year old) and it was really home-like. Their radio brought in carols from everywhere, including Winnipeg. At midnight we staggered back to Ward's cabin to hit the hay until daylight.

He also described school and life at Kelly Lake:

The school is nicely under way again but the old building is rather chilly on coldest days.... Did a little exploring today – out to Crooked Lake on snowshoes – six miles [10 km] from here. Had not been there before. Took about four hours there and back. I enjoy getting out into the air and snow after a week shut up in the school room. We had our "Xmas Celebration" for the school after New Year, because when we closed school for the holiday, none of the fathers and big brothers had returned from their trap lines. Am learning some Cree, but very slowly. It is hard to learn unless one lives right in their homes with them.

During the second half of the year, Andrews spent more time teaching geography and history to his older students. He incorporated local geography and history into his lessons to make them more relevant to his students. He later noted that his "life-long interest in the history of the Great Northwest began in this manner and about this time."

By the spring, he became restless again. "My time at Kelly Lake was coming to an end. Full and varied as it had been I began to think it was time for a change. Friends at Big Bar had written that my old school there was to be reactivated with an anticipated influx of new scholars." In May, he received a letter from Clarence Cover in Upper Big Bar, saying: "You can get your old job back again and we will be glad to have you. We were all glad to hear you would come back. I will look for you here any time you can make it and will be glad to shake your old fist once more."

Summer 1925

Before he left Kelly Lake, Andrews made plans with Jim Young for another summer adventure, this time with a larger party and more horses. In preparation for the trip Andrews learned how to "throw the diamond hitch" from Denny Cornock, homestead inspector for provincial Crown lands in the South Peace district. "Cornock's coaching in the 'diamond hitch' had practical application on our long trip to Jasper when we, too, became experts. I can still 'throw' it and take more pride in this qualification than in others confirmed impressively on framed parchments now hanging on the walls of my study."

The trip proposed for the summer of 1925 was on the Nose Mountain trail southeasterly to the Hinton area and from there along the wagon road west to Jasper. Young and Andrews would be accompanied by two of Young's friends, George White and Homer Wright. The four men would be guided by Sam Wilson, nicknamed Neestou (kinsman), a Metis member of the Shelter Band on Nose Creek. They would take 14 horses and two dogs. As he had done the previous two summers, Andrews kept a diary for much of the time.

On June 27 Andrews and Young travelled to Beaverlodge where they met Homer Wright. They picked up George White at his place the next day, and on June 29 the party reached the Wapiti River. Andrews described the part of the river they would have to cross as swift and about 100 metres wide in a deep valley about 60 metres below the upper benches. Because the river was too deep for fording, the men built a raft and with much difficulty got the horses to swim across.

They reached the foot of Nose Mountain on Canada Day. The next day they reached the top, where "both dogs plunged into a porcupine. Max is covered all over head and mouth. Thought we would have to shoot him, but at noon Jim and Neestou pulled all quills with pliers. They were an hour on this job." (Unfortunately Max disappeared a few days later.) For the rest of the day the party followed the trail along the top of a ridge that gave them a great view of the Rockies.

Nose Mountain.

In later notes Andrews wrote about the guide: "Sam had not been over this trail since boyhood, but in all cases of doubt he chose the proper lead." On the morning of July 3 "my horses and Jim's were all missing.... I went back about 3 miles to round up the culprits. Saw 3 moose. Got away at 7:45. Just after leaving camp, a bear with 2 cubs stampeded George's horses." By the end of the day the group reached the Kakwa River. In *Metis Outpost* (1985), Andrews' reflection on his pack trips of 1924 and 1925, he pointed out the places he visited that were noted in the Canadian Pacific Railway surveys of the 1870s. One of these was at the crossing of the Kakwa.

On July 5 the men crossed the Smoky River, which Andrews described as "about the same width as the Wapiti here, but swifter." Again, the men built a raft for themselves while the "horses swam with little trouble". In the evening, Neestou killed a young bull moose and prepared a special treat:

> He also brought in the moose horns which at this season were "in the velvet". Our curiosity about this was satisfied later, when after supper he placed them in the embers of the fire, and occasionally turned them over. Just before bedtime he pulled them out and showed us how to chip off the charred velvet to expose the soft, tender and delicious meat inside.
>
> This was a tasty late evening snack washed down with fresh sweet coffee.

Rain kept the group in camp the entire next day, so they "tried to eat up all the moose meat, but there was lots left." They set off again the following day and reached the forestry trail going through the area. On July 9 they arrived at Entrance, having journeyed more than 300 kilometres. There they met Ed Johnston, travelling alone from Edson with 10 horses and a dog to help herd them. Some of the men knew Johnston, so they invited him to join the group.

Rafting on the Smoky River.

Three days later the party reached Jasper where they spent a week selling most of the horses. Then the group dispersed. Andrews was going to Big Bar and joined Ed Johnston, who was heading west with three horses. Johnston travelled alone to Red Pass Junction, about 60 kilometres west, while Andrews remained in Jasper a few more days before taking the train to meet Johnston. From Red Pass Junction the two men followed the north bank of the Fraser River on the old railway construction tote roads to Tête Jaune Cache. With the assistance of a local person they crossed the Fraser on a dugout canoe while the horses swam the river. From Tête Jaune they followed the Canadian National Railway grade to Goat River, arriving there on July 28. A forest ranger and his wife gave them advice about the Goat River trail to Barkerville.

After four days of travel on the Goat River trail Andrews and Johnston crossed a summit overlooking a forested valley, with a high flat-topped mountain in the southwest. That evening they stopped on a side hill above Isaac Lake, today part of the popular Bowron Lakes canoeing circuit. "Followed till dark, no feed, so tied up. Windfalls bad and Devil's club thicker than hell. Isaac Lake beautiful, surrounded by high mountains on all sides, gravel bottom, water clear and green. Fish jumping like squirrels in evening. No room for beds, so rolled up in the trail, makes a good bunk!" In his notes Andrews wrote, "The beauty of the scene that night from our side hill bivouac high above the lake is still vivid. The moon above Mount Peever (7500 ft [2300 m]) was reflected in the lake far below."

The following day the two men rose early and reached the end of Isaac Lake by 9 am. "Good cabin with many names from Barkerville direction. Pulled out from meadow at 2 pm and reached west end of Indian Point

Lake. Here McKahe [a pioneer] was building a large log house. His wife gave us some nice lettuce and onions. They gave us information about trail to Bear [Bowron] Lake." Andrews and Johnston arrived at the lake in the early evening.

Andrews and Johnston arrived at Barkerville in the late morning of August 4. Andrews described the town in his diary:

> Took a roadway, lined on both sides with unpainted buildings of ancient vintage. All structures including sidewalks are built up on stilts. A creek used to run down through the town. The Masonic Hall stands out prominently, the date on it 1869. They say that the level of the town is 27 feet [8 m] higher than the original, due to build-up of tailings from hydraulic operations on Williams Creek.
>
> Barkerville is now quite dead, depending solely upon two operating mines, the Antler Creek Dredge and the Lougheed Hydraulic operation. Quite a few old prospectors in the vicinity. Some old boys 80 years mellow, but still hardworking hopeful gold seekers. Chinese make up a large part of the population.

Andrews and Johnston left Barkerville the next day and reached Quesnel two days later. In *Metis Outpost* Andrews summarized his two summer trips from Kelly Lake:

> This was my second arrival in Quesnel. It was a year before, almost to the day, that Jim Young, Fred Barber and I had arrived here in the old Ford car from Vanderhoof and Prince George. I had thus closed a complete loop from Kelly Lake – some 450 miles [725 km] via Pine Pass in 1924 and 563 miles [906 km] via Jasper in 1925, for a total of practically 1,000 miles. Ed and I enjoyed a couple of days at Quesnel in a pleasant camp, and the townsfolk were friendly. The time had come to settle our accounts and each go his separate way. Ed had proved a good companion.

Andrews also reported that years later he used aerial photographs to help him plot the routes of these trips in detail. He was even able to pinpoint each of the campsites he and his companions had made.

From Quesnel Andrews travelled to Big Bar where he met the school trustees and confirmed his arrangements for the coming school year. Then he went down to New Westminster to visit the Haggman family before beginning the 1925–26 school year.

Clarence Cover and some of the school boys.

Upper Big Bar, 1925–26 School Year

Andrews' cabin at the school was gone, so the first thing he had to do when he arrived at Upper Big Bar was find new accommodation. He stayed with the Cover family for a while, until he found a place of his own. "The Adlers had a small cabin which was served by an irrigation ditch for water and with a woodpile nearby. It had all the essentials for 'batching' and the rent was modest."

Eleven students attended Upper Big Bar school that year. Four came from Big Bar Mountain school which was closed. Andrews had taught only two of them in 1922–23. "When I had first arrived in Big Bar in 1922 at age 18, I had no experience as a teacher. In 1925, age 21, I had three years' experience.… I now smoked a pipe, was not averse to a snort of rum, and I did not lack confidence in supervising my school or addressing parents, local school trustees and others."

The inspector visited Upper Big Bar school soon after classes started. He reported that the trustees needed to make extensive repairs to the school and add more classroom equipment. He also noted that Andrews "made a good beginning with the work, and the outlook for a successful term's work is very promising". Now that Andrews was an experienced rural school teacher, his second year at Big Bar went smoothly, and he had more time to socialize. "I was among old friends and in an environment I loved." He also became friends with a person he had known at another location. "A happy coincidence at Big Bar in 1925–26 was renewing acquaintance with Herbert E. Drake, then teaching at Jesmond School only a few miles from

Herbert Drake.

mine. We had first met at Field (summer 1919). In the Big Bar country we enjoyed much in common." The two men did a lot of hiking together that year.

"When school closed for the summer in late June of 1926, I had saved enough money to apply to enter forestry at Toronto University." But Andrews first needed to upgrade his French and take analytical geometry, which he did at the University of Manitoba summer school. There would be no summer adventure this year.

During his four years as a rural teacher Andrews experienced his first professional employment and grew from a shy 18 year old to a confident young man. At both Big Bar and Kelly Lake he developed lasting friendships, while the summer pack trips gave him intimate knowledge of BC's geography and the importance of maps. In later years Andrews revisited the places where he taught, many of the locations he encountered during his summer travels, and some of the people he met during this part of his life.

Forester

The University of British Columbia had a forestry program that started after World War I, but the faculty was mainly engineers, because most of the logging in BC in the 1920s involved railways. There was only one full-time forester on staff and on his advice Gerry Andrews enrolled at the University of Toronto because it "had an established reputation in forestry" and "a full faculty of seasoned foresters". He began his forestry program in the fall of 1926 and graduated in May 1930.

Summer 1927

Gerry Andrews first worked in forestry during the summer of 1927, as one of ten members of the Manitoba Pulpwood Selection Survey in the previously uncharted Precambrian Shield east of Lake Winnipeg. At that time the federal government still had jurisdiction over the natural resources of the three Prairie provinces and the project was sponsored by the federal forestry department and the Manitoba Paper Company. The crew that Andrews worked on was one of at least ten deployed in the Lake Winnipeg area that summer.

> Our specific assignment was the Poplar R. watershed, then unmapped and largely unknown. The strategy was to start at its headwaters and work, generally, downstream to its mouth, some 50 miles [80 km] north of Berens R.... For transport we had four excellent "Chestnut Prospectors" model 17-foot canoes, muscle powered. Whether outboards would have been better is debatable. Fuel for the former was grub, for the latter, gasoline. Either had to be brought in from outside.... We became expert canoe men, up and down rivers and on the big lake with its formidable wind and waves.

Canoes on Wrong Lake.

Andrews observed that they were supposed to have air photos but they were not done until that summer and did not reach them until Poplar River Mission on August 10 "when they were only of post mortem interest, explaining many of our troubles and wasted effort which they would have avoided." This was Andrews' first exposure to aerial photography and the beginning of his interest in using this technology to improve forestry techniques. The crew chief was Herb DeVeber, a person all the crew members respected. The cook was Alf Coles:

> When I first saw him [Coles] at Pine Falls late May 1927, dressed in immaculate white, I thought he was an Anglican priest. He maintained this décor in the wilderness. He fed us with yeast bread instead of bannock made with baking powder. On portages he insisted on being put ashore carrying his yeast culture in a pail, like a treasure.... It was a healthy life, lots of exercise, paddling to work and on moves, carrying loads over portages and making camps.

The crew members picked wild berries and ate fresh fish which were abundant. "On rainy days and 'Sundays' we played cards, mostly bridge, and had discussions about religion, philosophy, etc."

Andrews' diary for the summer describes the work and the rugged travel through the area. He left Winnipeg on May 25 and travelled by steamer up Lake Winnipeg to Pine Falls. While the crew was there for three days they toured the mill. On May 28 they left on the steamer *Phyllis Williams*, arriving at Berens River in the evening. The crew spent three days there preparing for their canoe trip up the river. During that time Andrews attended a Sunday service at the mission church and saw two float planes arrive. It took a week for Andrews' crew to paddle up the Berens River waterway to Little Grand Rapids. "The Berens River, some 110 route miles [180 km] and 52 portages to LGR was well used, well known and probably had been roughly surveyed." The route through the waterways was occasionally confusing and a few times they had to backtrack to find the right channel. In his diary for June 3 Andrews noted: "Two canoes with five Indians, passed going up-stream in pm on other side of canyon. They say that side better. They really know how to portage."

On June 10 the crew left Little Grand Rapids, searching for the divide that would take them into the headwaters of the Poplar River where the men were to begin their forestry work. The following day Andrews wrote: "Cold and more rain. Bucked high NW wind all day up fishing Lake.... Reached Pungachic Indian village 11:30. Willie Bushie and his partners showed us the short portage into headwaters of Stony Cr. [Assinipan River, a tributary of the Poplar]. Camped on island at NW end [of the lake]. The Indians could not speak English, but I did fairly well with my Cree." The Cree that Andrews had learned at Kelly Lake proved useful in finding the route for the crew.

After a day's rest the crew began their forest inventory and Andrews got a practical introduction to his work. "Spent morning learning how to run strip and estimate. Observations on trees and flowers, mosses, birds." The infamous mosquitoes and black flies of the Precambrian Shield caused trouble for the men and Andrews' diary contains many entries like this one, from June 15: "Working on strips, flies terrible. Home at 8 pm, tired, disgusted and eaten by flies. Slept less than 2 hrs. Bites on chest worse than anything I remember." And a few days later: "should bathe but hate to brave the mosquitoes."

By June 25 the crew reached Wrong Lake, where they spent almost two weeks. Andrews noted their celebration of Dominion Day. "Did nothing but smoke, eat, shiver, discuss religious philosophy, speculate what folks are doing back home on Canada's Diamond Jubilee." In later notes he added: "An interesting item, not mentioned in my diary, was finding an old iron bullet mould in the ashes of an ancient campfire, on the island where we camped at Wrong (Drunken) Lake, late June. Likely it was a trade item in the early days of the HBC store at LGR."

On July 7 the crew left Wrong Lake and headed up the main Poplar River. They were running low on food and expected either a plane or packers to bring in supplies. By July 19 Andrews noted: "Grub now bread, beans and tea." The men also realized that they were probably not on the main Poplar River. "Think we are not on the main branch, shown on the map draining Shallow L., but on the north branch indicated as entering Rice Lake from the NE." Three days later two crew members found their food cache but by July 27 they were out of food again and awaiting the August supplies.

When packers arrived with their food in early August, they told the crew that their supply plane had crashed, killing the pilot and his two passengers. They also brought orders for the crew to proceed down to the mouth of the Poplar River.

A bark house at Little Grand Rapids.

On August 10 the crew reached the Poplar River Mission, where they witnessed the arrival of the government treaty party the next day:

> HMCS *Bradbury*, Capt. Brian, steams in at 7:30 am with Treaty Party: Stevenson, Indian Agent; Millich, clerk; Bertram, RC Priest; Coldwell, Methodist Priest; Miles Mitchell, Indian Chief; Cubby, counsellor and Dr Gibbs. Went down this pm to see doings at Mission, 32 Indian tents, treaty money to 107 souls. Philip Thomas interpreter. Dance.

The crew spent the remainder of the summer season working along the northern part of Lake Winnipeg. In his diary Andrews expressed his interest in local history. When the crew arrived at Grand Rapids on August 20, he described following the rapids to the upper end where he found the remains of an old Hudson's Bay Company post. On his return trip he walked along the HBC railway ("about 70 years old") for almost eight kilometres. He also noted that there was an Anglican mission and school, and that the mission began about 75 years ago.

On September 15 the crew bunked at Warren Landing:

> Last night had the surprise of the whole summer! Had just got to bed, when Frenchy stepped in and shouted "All out boys! Save your canoes, the flood is carrying them away!" We rolled out to find bunkhouse battered by waves and surrounded by water. Rescued canoes. Rupe ferried Mrs Cook and her kids to log house on higher ground, her own cabin in 2 ft of water.... We lend the local folk one canoe and have three ready to move to a ridge west of settlement. At midnight wind swings to west which relieves our crisis. Flood recedes slightly, so we return to our bunks at 1 am. I could not sleep....

However the west wind makes it bad for the people on the east side
of river. There, tourist shack washed down against wharf, cook house off
level – whole settlement partially submerged. We can't cross over to help
due to high seas and strong current.

The storm caused severe damage, but fortunately no injuries.

By September 21 Andrews was back in Winnipeg, ready to return to
Toronto for the second year of his forestry program.

Summer 1928

Andrews' summer employment in 1928 gave him quite a different experience in forestry. That year he worked for the Laurentide Power and Paper
Company at its forestry station near Grand Mere, Quebec. There Andrews
learned about forestry nursery practice, tree plantations and their management, and simple land surveys. His French improved considerably that summer, because his fellow workers would not speak English.

The chief forester for Laurentide was. "forester extraordinaire" Ellwood
Wilson, one of the leaders in this field in Canada during the first decades of
the 20th century. Wilson was an innovator in many aspects of forestry work
and was particularly interested in utilizing airplanes to improve techniques.
In 1919 he was able to get two war
surplus planes on loan from the federal government and used them that
summer for fire spotting, for transporting employees, provisions and
equipment, and for aerial photography. The Canadian Air Board
described this as "the first practical
use made of civil aviation in Canada". In 1922 Wilson met Sherman
Fairchild, an American who had
just developed a new commercial
air camera. Soon afterward Wilson
founded Fairchild Aerial Surveys of
Canada (renamed Fairchild Aviation
in 1926) to do aerial photography
in Canada. Fairchild also produced
airplanes that were used for general aviation purposes but were particularly suited for aerial photography. Wilson was a mentor to many

Ellwood Wilson.
(Pierre Thiffault collection.)

foresters throughout Canada. In the 1930s he and Andrews communicated frequently, and Wilson was a vital supporter of Andrews' efforts to incorporate aerial photography into forestry practices in BC.

Summer 1929

In 1929 Andrews returned to BC, working for the provincial government's Forest Branch (later called the BC Forest Service) as a timber cruiser on the Elk Forest Survey in the southeastern part of the province. This survey was typical of forest inventories conducted in BC in the 1920s. When the Forest Branch was established in 1912, one of its mandates was to develop an inventory of the province's resources. In the initial years the inventories were mainly a general reconnaissance done by one or two men. Before World War I these men sometimes travelled with the government survey parties working around the province. After the war, forest inventories became more detailed and scientific.

In 1925 Frederick Davison Mulholland became chief of Forest Surveys for the province. Mulholland, one of the important early foresters in BC, developed guidelines so that all forest inventories were done in similar fashion. For the inventory a simple survey line called a base line was established through an area of the forest. At mile intervals perpendicular strip lines were used as a basis for mapping forest cover and topography. Andrews described the process this way:

> The usual procedure was called a 1¼ per cent cruise, covering the area by
> a grid of examination strips on the ground spaced a mile apart. Direc-
> tion was by compass, distance by rough chaining, and elevation carried by
> abney level. The strips were tied directly or indirectly to existing survey
> control such as district lot corners, traverse and triangulation stations.

Tallies of mature timber were made in a band one chain (66 feet or 20 metres) wide, and a narrower strip in young timber, for volumes, species, density and age-height measurements at intervals. Since there are 80 chains in a mile, 1/80 or 1¼ per cent of the timber was being sampled. From this sampling the volume of timber in between the strip lines was calculated. If there was a change in species or age of trees along the strip line this was noted. The standard day's work was a strip two and a half miles (4 km) long. Occasionally crew members would hike onto mountain tops or high places to sketch the countryside and record any significant geographical features that were not included in the samples. Andrews wrote: "Between and beyond the strips, topography and cover were sketched, and the 100-foot contours were mapped in as crossed. The result was a realistic if approximate topographic map, with forest cover types."

During the last half of the 1920s several crews went out each summer to conduct forest inventories throughout the province. The crews consisted of a party chief, an assistant party chief (on large crews), compass-men who did the necessary surveying and laid out the strip lines used for taking the sample of trees, cruisers who measured the height and diameter of the trees and occasionally checked for the age of a stand of trees with an increment borer, and a draughtsman who normally stayed in camp and did the mapping and calculations in the office tent. Each crew also had a cook and usually a packer. Most crews had at least 10 people who spent about four months in the field. These inventories provided great detail about the conditions in a particular forest in the province but they were labour intensive and covered only a small area of the province's vast forests each year.

The main purpose of the forest inventory was to provide the best possible estimate of the quantity and quality of timber in a forest, but Mulholland was also interested in having foresters report on the conditions they found as they travelled through the area – soil, wildlife, water, road access, trail conditions, other economic uses of the forest, etc. He was concerned not only with the economic value of the timber but also with ensuring that the province's forests would be an economic asset in the future. In a 1931 article for the *Forestry Chronicle*, Mulholland wrote:

> A timber cruise, however detailed and thorough, is simply a crop survey: how much wood? how and at what cost can it be moved to the manufacturing plants? A forest survey is an economic survey, concerned with all branches of forestry: policy management, silviculture, regulation of the cut, utilization, protection. Its ultimate object is to provide information to enable a forest to be administered for permanent wood production.

In their biography of Frederick Mulholland, foresters Gerry Burch and John Parminter described him as "the father of sustained yield forestry in British Columbia".

After three years of studying forestry at the University of Toronto, Andrews was happy to return to BC for a summer. Travel to the job site included enjoyable boat trips, first aboard the SS *Minto* on the Arrow Lakes and then the SS *Nasookin* on Kootenay Lake: "The relaxed luxury of the old lake steamers and the delights of passing scenery from their decks persist in happy memory."

Hugh Hodgins was the party chief, and the crew started their work near Yahk.

> After about a week in this location the Chief was horrified to learn that what appeared to be a farm dwelling in a field across the river was in truth a "house of entertainment" run by a family of negresses under supervision of "Auntie", an older relative. The clientele for this establishment was drawn from railway employees and tie hacks up Hawkins Creek

The crew aboard the Nasookin: standing (left to right), Davis Carey, Bob Well-
wood, unknown, Andrew Gordon (leaning on the steps); seated (left to right),
The crew members included: (seated, left to right) George Cornwall, Harold
Mahon, Ian Mcqueen, unknown.

to the east, on the old CPR Tie Reserve. However the Chief ran no risk
of contributing to juvenile delinquency, due to a combination of youthful
innocence and physical exhaustion from our strenuous daily labours.

Andrews soon learned that the daily routine of forestry work was difficult in BC:

En route from Victoria to the Elk job I well remember eyeing the steep
mountain slopes all round and wondering apprehensively if the cruise
strips would have to be run up their forested declivities, recalling the more
bland topography in Manitoba and Quebec. My fears were soon confirmed. In fact our strips were laid out, seemingly to cross the maximum
number of contours, in both length and direction. Eventually we became
hardened to this feature, and from all appearances, survived.

Later in the summer Andrews was part of a small group that was sent to
work in the Wigwam River valley in the southeast corner of the Elk Forest.
The men ran a traverse up the valley bottom with lateral cruises every mile
that started from the valley and went up the sides of the adjacent mountains.
"As we had to map topography and drainage as well as timber types, much
had to be done by sketching, using compass bearings and judging distances.
The far ends of our strips were often planned to afford a known position
high above timber line for sketching a wide vista of country."

Packer Joe Falconer fording the Wigwam River.

One day while having lunch on top of a ridge with compassman Davis Carey, Andrews walked by himself to a viewpoint a short distance away where he could sketch the detail of a tributary valley. Years later he described his unexpected encounter in an article called "My Bear":

> I followed just below the ridge for shelter from a brisk west wind. There were still patches of snow. I noticed rocks had been freshly turned over, probably by a bear looking for ants and bugs. A few moments later I looked up and there he was, about 50 feet ahead and just above. I was surprised that he did not retreat, but stood facing me. My next surprise was that it was a grizzly and a big one.
>
> [Rather than panic I] decided to detour around below him. When I moved sideways he advanced, so I stopped and he stopped. After a couple of such moves, he was within 25 feet, and evidently agitated, head down and sort of whining. I talked to him, like to a strange dog, but decided, if he attacked, I would give him a whack on his snout with my inky borer that he would not forget.

Then Andrews tried a different tactic:

> Instinctively I stepped backward. He did not move. I took another step, and another. He stayed put. Gradually, still facing him, I dropped down in a depression and the ground came between us. I then retreated smartly, watching that he did not follow. When I reached Davis, he said "Gee you're white, what happened?" I said, "There is a grizzly over there – let's get the hell out of here!"

Like Kelly Lake, Andrews witnessed the transport of illegal alcohol that summer:

> Rum running was active down the Kootenay valley in 1929 with high-powered black Hudson sedans favoured for this traffic. They tore down the

road at high speed, loaded to the roof with cases of whiskey, trailing huge white clouds of dust. Between Grasmere and Roosville they would disperse on side trails in the open park-like yellow pine forest to rendezvous with their contacts from south of the line.

Late in the summer several large forest fires in the East Kootenays caused the crew some problems: "smoke from increasing forest fires began to burden the air and our work was interrupted by having to fight a fire which started far up Gold Creek." The men eventually returned to their work, continuing until mid September. The remainder of the survey of the Elk Forest was completed the following summer. In summarizing the summer of 1929 Andrews wrote: "Besides the inspiring scenery of the East Kootenays, the wealth of forest flora stands out.... About mid September I boarded the CPR at Elko, eastbound for my final year at Toronto. The steep mountain slopes conquered, no longer so formidable, were left behind."

That same summer the BC government started a large project called the PGE Resources Survey that would have an impact on Andrews' career. In an effort to provide assistance for the Pacific Great Eastern Railway to complete their line to Prince George and eventually continue into the Peace River region, this survey made a thorough examination of four blocks of land, the largest being in the Peace River area. The government hired surveyors, foresters, geologists and other professionals to report on conditions and economic potential in the land designated for this survey. The PGE Resources Survey was the government's first large project where aerial photography was used. This was done by the RCAF and Western Canada Airlines. The RCAF used two Fairchild 71 aircraft.

In the 1920s aerial photography started to be employed in eastern Canada and the Prairies for mapping and examining resources of an area. But it had not yet been used to any extent in the mountainous terrain of BC. For the PGE Resource Survey the surveyors were to do their traditional triangulation surveying but not in as much detail as usual. Main control points were surveyed and wooden tripods wrapped in white flagging cloth were placed over them. These flagged markers, with their fixed survey information, would be visible in the aerial photographs. Since the distance and direction between each survey marker was known, maps could be made from the aerial photographs. These were then checked against the information gathered by the surveys. As is often done in the implementation of new technology, the traditional method and the new method were used simultaneously to ensure that the new technology was accurate.

Andrews later wrote: "The famous PGE Resources Survey ... sparked the first big air survey effort in BC.... It was not only a convincing demonstration of the potentialities of air photography in mapping, but it familiarized many of BC's survey and engineering personnel with the new method

by actual experience." Although there were problems with the aerial photography, the information gathered showed that photographs were able to
provide more detail and cover a larger area more quickly than traditional
surveying. Photogrammetry, the use of aerial photographs, had made its start
in British Columbia.

Summer 1930

In the winter of 1930, as Andrews prepared to graduate from the University
of Toronto, the Depression grew worse in Canada, and he wondered about
opportunities to obtain a position in forestry. He particularly wanted to return to BC, and wrote to Mulholland regarding a position with the Forest
Branch. At the end of February Mulholland replied:

> There are not at present any permanent vacancies into which we could
> place you and it is impossible to say yet what there may be during the
> present year. With regard to temporary work, our plans are not yet com-
> pleted, but I think there is very little doubt that we will be able to offer
> you work either in the Forest Surveys division or Research division.

By mid April Andrews had not heard further from Mulholland so
he wrote another letter. "If you cannot give me work this year, I should
appreciate it if you would wire me at my expense. I have refrained from
applying for a position elsewhere till hearing from you definitely. As you
probably realize, most of the openings here for graduates are being rapidly
filled by my classmates." But Mulholland had already written to Andrews on
April 10:

> This year we are dividing the work east of the Kootenay River into two
> parties and want you to take one of them in the Flathead River watershed,
> doing the same class of work you were engaged on last year....This work
> will be continuous through the winter as we will require you to come to
> Victoria on completion of the field work and take care of the reporting
> and mapping of your work in the office.

Hugh Hodgins would finish the Elk Forest survey while Andrews would
do the Flathead Forest to the east. The work would start in mid May and
Andrews could join the crew on June 1 after his final exams. Andrews was
ecstatic. Mulholland had not only offered him employment, but it was a
supervisory position that would extend through the winter. He immediately replied to Mulholland on April 17:

> Your letter of April 10th, received today, renders my letter of April 16th
> not in order. I shall be very glad to have the opportunity of working for
> the BC Forest Branch in the manner outlined in your last letter. I realize
> that the work planned for me carries with it a pretty big responsibility,

rather bigger than I should have anticipated. I only hope that I can "fill the bill".

On June 1 Andrews stepped off the train in Corbin, a small coal-mining town just west of Crowsnest Pass, and walked about 25 kilometres, crossing snow-covered Flathead Pass, to the Flathead town site where he joined his crew. "I arrived at camp with blistered feet, tender from sedentary months at university." Mulholland's assistant, Gerry McKee, had started the survey. Among Andrews' crew members that summer was Lorne Swannell, son of the well-known BC land surveyor, Frank Swannell. Lorne wanted to be a surveyor like his father, but Frank did not approve, so Lorne had gone into forestry instead. Gerry Andrews and Lorne Swannell became life-long friends – longer than most people, because both men lived for more than a century – and along the way each man rose to the top position in BC in his field of work (Swannell as chief forester and Andrews as surveyor general). Through Lorne, Andrews met his father, Frank Swannell, who became a mentor in Andrews' surveying career and also a friend for life, and his younger brother, Art, who worked with Andrews in surveying after World War II. Lorne kept a diary for the 1930 field season and took photographs, which would become a valuable record.

The Flathead River valley, in the southeast corner of the province, is bounded by the USA on the south and Alberta on the east. The entire region is above 1200 metres elevation, with large sections of subalpine and alpine terrain, and several high peaks going up to about 3000 metres. It has minimal agricultural potential. In addition to the forest inventory, Andrews was asked to make his recommendations regarding a national park for the Akamina and Kishinena area on the east side of the Flathead River valley. With Waterton National Park on the eastern boundary and Glacier National Park, in Montana, to the south, there had been considerable interest for several years in having this portion of the Flathead drainage designated as a park.

In 1930 there were only a few people in the Flathead valley: trappers and some small crews working on the oil wells at Sage and Akamina creeks. One of the trappers was Charlie Wise, who lived in the Flathead for many years, initially in Montana, and then in BC after 1919. When Andrews asked about trails in the area, Wise replied: "Oh, I don't pay no 'tention, I just go anywhere." A Canada Customs officer and his family lived at the border crossing. There was a road from Waterton Lakes National Park in Alberta to the oil wells on upper Akamina Creek, and another from the oil wells on Sage Creek down into the United States.

> The plan [according to Andrews] was to work from the upper watershed above townsite down the main valley, southward to the Montana border, doing lateral valleys en route, then work easterly up the Kishinena

A well-used First Nations transportation route from the Rocky Mountains to the Tobacco Plains area near the Kootenay River crossed the Flathead River valley in the area where Andrews' crew worked. Although First Peoples stopped using the route before 1930, the men found tepee frames and other evidence of their campsites at a few locations.

valley to Akamina Creek and Pass, and finally to go out over the pass to Waterton Lake, in Alberta. Camp was moved about every two weeks with some 15 pack horses owned and operated by Jack Aye.

On the south Andrews tied the Flathead Forest survey into survey monuments along the Canada-United States boundary, and on the east into stations on the BC-Alberta boundary. On the west he used an unusual survey done by John McLatchie in 1904. McLatchie surveyed the east boundary of Lot 4589, a huge coal reserve for the BC Southern Railway. Beginning at the highest point of the Railway in Crowsnest Pass, the line ran due south for more than 43 miles (70 km) to the international boundary. At each mile McLatchie placed a huge squared wooden post set in a mound of earth or a pile of rocks. Although there had been a fire in the area in 1929, Andrews was able to locate several of McLatchie's mile posts.

The men worked whenever there was good weather and took their "Sunday" when it rained. Swannell liked to listen to music and persuaded the crew to contribute four dollars each toward the purchase of a portable gramophone, which would go by a draw to one of them at the end of the season. The gramophone became part of the entertainment on Sundays and some evenings.

In mid June Mulholland came to visit Andrews, check on the survey, and take Gerry McKee back to Victoria. Then the crew moved into the

Oil well at Sage Creek.
(Lorne Swannell family collection.)

southern part of the Flathead River valley. They established a camp at a place called Butts, named after a family that had once lived there. The crew spent about a month at Butts because it was centrally situated to several valleys that drained into the Flathead.

At the end of July, Swannell reported in his diary that the crew cook, John Clement, had "cut his leg badly with axe, [and we] brought him over to get an oil well man to take him by car to [Belton] Montana to be stitched up." Andrews noted later that this was the only serious accident he encountered during his years of forestry field work, and it happened because the man "was nudged by a horse and cut his thigh with his own sharp axe.... About a week later we learned that John was ready to return. With a couple of horses from Jack's string I went down to fetch him."

In mid August Andrews' crew moved down the Flathead not far from the US border. Mulholland returned to visit Andrews again for a few days. Though Mulholland was one of the highest ranking foresters in the province, he still liked to socialize with the crews in the field. Swannell wrote that one night on this visit, they played bridge and on another evening "after supper, long discussion with FDM, mostly sociological – intermarriage of races – Mussolini – Russia, etc. Interesting." Since the crew was going to work up the Kishinena drainage toward the Alberta border, Mulholland approved Andrews' plan to finish the season by going out through Waterton Lakes National Park.

A few days after Mulholland left, the men had a party. In his diary Swannell commented:

> To my disgust, GSA entrusts me with the Scotch which we got out of camp without Johnnie seeing it. Hell of a good party. Jack & Sam Mc-Donald arrive at 10, having come 55 miles [90 km] in a day. Brought Manhattan cocktails, liven up the party.... Try square dancing with Jack acting as caller – nine records broken in consequence. Home at 4:30 am.

Mount Gerry Andrews

Trachyte Ridge, with Mount Gerry Andrews in the centre.

In March 2011 the BC government officially named the highest of three small peaks on Trachyte Ridge Mount Gerry Andrews. It is on the west side of the Flathead River and has a commanding view of the valley south into the United States as well as the surrounding mountains. Although Andrews' survey focused on the area from the Flathead east to the Rocky Mountains, he occasionally climbed mountains on the west side that provided a clear view across the valley. On June 29 he climbed Trachyte Ridge (called Lookout Peak in his notes) and took a panorama photograph east across the Flathead River. On July 6 Andrews and Lorne Swannell climbed a peak near their camp on the east side of the valley and looked across to the mountain that now bears Andrews' name. Swannell described this day in his diary: "GSA & LFS up at 6, make own brekka [breakfast]. Climb to top 8000 ft [2500 m] mtn. Best mtn scenery I've ever seen, saw oil wells on Sage Cr. directly below. With binocs could see roads & houses in Montana. Lyall's larch near summit. GSA took 16 photos. Quick trip down, swim in river."

Mount Gerry Andrews is one of five features in the area named for BC land surveyors.

There was excitement next morning in camp. "Self & GSA only ones up for brekka. Fire breaks out in hdwtrs [headwaters] Couldrey Cr. GSA gets Stan & me to go up mtn to locate it, he goes down to States to phone rangers. Returns, says Americans already on it, located at 7 am and had men on it by 11. Some system and real cooperation." And the following day, "GSA & I go up to fire, Americans have it pretty well controlled."

In mid September the crew travelled to upper Akamina Creek. Lorne wrote: "Move up to Akamina Oil camp. I help Jack pack. Get there, all bldgs locked, so smash couple of padlocks to enter. A1 layout, new cabins, spring mattresses. Johnnie nervous about breaking in, won't use bed in cook house, sleeps on the floor." On the next day: "Take Sunday. Another rainy day. Stan & Ian in early. Caretaker comes & sanctions our presence. Johnnie relaxes. Bridge & poker. Surprise, I win."

Later, Andrews wrote about the time they spent at the Akamina camp: "We were glad of indoor accommodation because snow flurries had begun. Hazard, the lonely watchman, enjoyed our company, was most co-operative and of course was our guest at meals." There was one drawback to their living quarters. "The bunkhouse at Akamina had a lumber ceiling. In the space above, after we retired, nocturnal packrats, jumping over the stringers, kept us awake. They seemed to be playing tag. This annoyance was partly remedied by firing a .22 rifle into the ceiling which seemed to quiet them. I don't remember paying damages for the bullet holes."

On September 20 Swannell wrote in his diary that the crew had run a strip that would be the farthest east in the province. By the end of the third week in September, Andrews' crew had completed the inventory for the Flathead forest and came out to Waterton. From there Andrews went by car with Jack Aye to Elko where he joined Hugh Hodgins and his crew, who were completing the Elk Forest survey. Andrews worked with Hugh until the survey was completed. "Returning by Victoria by road with Hodgins' party in several vintage vehicles, our first overnight was at the old Davenport Hotel in Spokane.... Second night we stopped early for a tire repair at Ellensburg.... In those days there was no passable road from the Interior to the coast in BC so our route was to Seattle and by ferry from there to Victoria." Andrews had successfully completed his first season as party chief of a BC Forest Branch crew.

A Proposal for a National Park

From the Akamina camp Andrews spent some time exploring this area near the borders between BC, Alberta and the USA, examining its potential for a park. He hiked to the top of Mount Festubert along the Continental Divide and visited Akamina Ridge near the US border. Swannell describes one trip. "Do siwash over to Kintla – by going to beat hell & getting an early start get back by night – at least get back to camp two hours after dark. Kintla quite a scenic place – not pretty but wild bluffs. Came back by way of Wall Lake – well named – surrounded on 3 sides by 2000-foot bluffs. Saw seven goats and four black-tailed deer."

Wall Lake from Akamina Ridge. Today this is part of Akamina-Kishenina Provincial Park.

In his government report Andrews wrote an appendix called, "The Proposed Kishinena Park":

> The area recommended as most suitable for park purposes is that part of the Flathead Forest occupied by the watersheds of upper Kishinena Creek, Starvation Creek and Kintla Creek. This region constitutes a wedge shaped tract of approximately 79 square miles [205 km²], which is bounded by Waterton Lakes [National] Park on the north east and by Glacier National Park on the south in the state of Montana. It is thus bounded on two of three sides by established parks and game preserves in Alberta and Montana. It constitutes a topographical unit separate more or less from the main body of the Forest. A high proportion of rugged mountainous country of low forest productive value renders this area of least value for inclusion in the Flathead Forest for the purpose of growing timber and at the same time makes it admirably suited for dedication to park purposes.

Despite Andrews' recommendation and other proposals over many years, it was not until the 1990s that the BC government created Akamina Provincial Park. It includes most of the area recommended by Andrews.

1931

In 1883, during construction of the Canadian Pacific Railway, British Columbia gave the federal government a land grant of 20 miles (32 km) on either side of the railway called the Railway Belt. The province also transferred a large section of land in the Peace River known as the Peace River Block. In 1930 this land reverted to the BC government. Although the Depression caused economic conditions to become worse and the BC government was cutting back on its expenses, the Forest Branch wanted to do inventory surveys on the Railway Belt land. The Forest Branch's government report stated: "A forest survey was commenced of the National Forests in the Railway Belt which were transferred to the province last year in order to ascertain their comparative economic forest value." This was an important economic resource that had easy access to transportation. For the summer of 1931 Andrews was placed in charge of a large crew that was responsible for doing an inventory of the Tranquille and Niskonlith forests in the Kamloops area. Lorne Swannell was a member of his crew again.

This summer Andrews finally had an opportunity to use aerial photogrammetry in his forestry work.

Aerial photography started in the 19th century when people took pictures from balloons. The pictures provided a different and broader perspective of the earth. After the invention of the airplane, the use of aerial photography increased rapidly, particularly during World War I. Better airplanes and improved cameras enabled military personnel to employ aerial photography for a variety of purposes.

After the war, photogrammetry – using photographs in mapping and surveying – progressed as scientists developed mathematical formulas and procedures to obtain precise information from aerial photographs. Andrews wrote: "Photogrammetry, in its original sense, could be narrowly defined as the art and science of deriving quantitative information about things whose images could be recognized in a photograph, and about their relative position in the external environment which the photograph covers."

Aerial photogrammetry required an airplane, camera, film and personnel to produce, process and interpret the images. To obtain reliable data the airplane had to be flown at a fixed speed and altitude over precise flight lines to obtain a uniform set of pictures, the camera had to produce high-quality photographs, the camera operator had to take the photographs at precise intervals so each one had a similar amount of overlap with those taken before and afterward, and the film needed to provide clear resolution.

A pair of overlapping aerial photographs, placed side by side, would appear as a single three-dimensional image when viewed through a stereoscope. By knowing the altitude at which the pictures were taken and the

Andrews' crew in 1931: standing (left to right), Gerry Andrews, Marc Gormely, Alf Buckland, Bill Hall, Bill Ingram, Larry McMullan, Lyle Trorey, Alex Gordon, W. Hays; sitting (left to right), Bob Wellwood, Bob Anderson, Lorne Swannell, Jimmy Brown, Clark McBride, Mickey Pogue.

focal length of the camera, the scale of the aerial photographs could be determined. From the scale, distances could be calculated and the pictures could be used for mapping and obtaining other information.

During the 1920s technological developments improved aerial photogrammetry and it began to be used in areas of Canada where the topography is relatively flat. Andrews described this process:

> By the mid 1920s sizeable tracts of flattish country were covered with oblique air photography; i.e. the cameras tilted obliquely to include the horizon image along the top of the pictures.... Forest cover for pulpwood and saw timber were mapped in this way, and geological interpretations gave great promise. The technique became widely known as the Canadian Oblique Method.

The mountainous terrain of British Columbia had very different conditions. The PGE Resources Survey had shown that aerial photogrammetry could be used for mapping in the province. Could it also be used in forest inventories?

Andrews talked about his first experience with aerial photographs in an article by Charles Lillard published in *Fores Talk* magazine:

> "My preparations were well advanced for taking to the field when a pal, W.A.A. (Lex) Johnston, who worked … on forest reconnaissance, informed me that part of the Niskonlith Forest was covered with 1928–30 RCAF air photos, which were "on hand" in custody of Major Alistair I. Robertson, BCLS, then air-photo librarian for the surveyor general, with

An early mapping stereoscope.

whom I was in "bon accord". The photos were discreetly "borrowed" for the summer, without reference to higher authority." Gerry had a wooden box made to protect them. A stereoscope for observing the photos in three dimensions was also scrounged.

The crew began the season working for a week in the Clinton area and then spent the first half of the summer in the Tranquille Forest, so there was no opportunity to use the aerial photographs until late in the season.

In a three-part article on the history of British Columbia's air surveying that appeared in the *BC Historical News*, Andrews wrote:

> By good luck, two members of my party, Marc W. Gormely and William Hall, had worked with Norman Stewart, in the winter of 1929–30, plotting photos for the PGE Resources Survey. With their help we contrived to make radial plots of the photo-strips on waxed (lunch) paper, far from ideal. We soon found, even with the photos so crudely plotted, how the most efficient layout of our ground examination strips could be made, both for tying them in on the map, and for best sampling of the various forest types. It became reassuring, also, that where our ground strips crossed the various features to be mapped, as identified and located from the photos, there was good agreement. For the terrain between ground strips, a mile apart, the photos offered infinitely more detail than orthodox interpolation by observation, guessing and sketching.

Andrews described what happened in the field to Charles Lillard:

> On one wet weekend at the Louis Creek Ranger Station the photos were spread out across the cabin floor and the men began to study them. "Just as we were in the middle of this chaos the boss arrived. He had the habit of not letting you know he was coming. He just appeared." The boss walked in, looked around, and said, "What the hell is going on here?" So Gerry had to explain that he wanted to study the usefulness of using aerial

Truck trouble on
Tranquille Road.
(BC Archives NA-10821.)

photography in surveying. "The only way to map the country is to get out and see it," he [the boss] said.

But Gerry stood his ground. "I was convinced we had something." Finally, the boss allowed them to use the pictures but only if Gerry "would work as though you didn't have them". The boss went on to say that the men would have to plot the photos in their spare time.

Gerry added, "He knew there was no spare time on a survey."

In his report at the end of the season, Andrews demonstrated how he had been able to use aerial photogrammetry successfully in his forest inventory of part of the Niskonlith Forest:

In preparing the final returns of survey after the field season, in addition to the customary maps, I made a special sheet for the area covered by air photos, which showed, with a suitable legend, the information as obtained with the photos and that which would have accrued without them, by ground methods alone. The result was convincing proof of the virtues of air photos. The extra detail and the subtleties of outline derived from them were striking, and final proof was that wherever our ground strips and traverses crossed features from the photos, there was full agreement. Mulholland was satisfied, and thereafter became a staunch and influential supporter of air survey.

1932

Now that he had demonstrated the value of aerial photography for forest inventory in BC, Andrews wanted to learn more and improve his knowledge in this field. Over the winter he wrote to Ellwood Wilson, the forester he had met at Laurentide in 1928.

Wash day.

In connection with the survey of the Niskonlith and Shuswap forests in the former Dominion Railway Belt in the province, which was
commenced last year, we have for the first time had occasion to use in a
systematic way aerial photographs which were available for a greater part
of the area covered. It so happens that no member of the staff of our division had had previous experience in the application of this type of data to
forest reconnaissance surveys. Since I was assigned to the direction of the
field party covering the area last year, and to the compilation of the maps
and report in the office during the present winter, it has been my concern
to gain a working knowledge of the application of aerial methods to our
type of work....

In March, on Wilson's recommendation, Andrews wrote a similar letter
to Stuart Moir of Fairchild Aerial Surveys in Dallas, Texas.

As the 1932 field season approached, the continuing Depression meant
even more cutbacks in government funding. Fortunately for Andrews the
Forest Branch continued its inventory of the forests that were in the Railway Belt and he was put in charge of a survey adjoining the Niskonlith.
He proposed to Mulholland, "to reduce costs and to facilitate operations, a
preliminary map of the Shuswap Forest (east of the Niskonlith) be compiled
from air photos available, prior to going to the field." Mulholland agreed,
so Andrews proceeded. "This was done with some temporary help, better
facilities in the Victoria office, and experience gained on the Niskonlith job.
With this map, the air photos, a stereoscope and the usual kit of instruments,
I set out for the Shuswap Forest with a single technical assistant, Stan G.

Gerry Andrews using
an abney on Crowfoot
Mountain.

Bruce." (Bruce had been on his 1930 Flathead crew.) Andrews also hired a
packer for the first part of the summer

> for areas too remote for access by canoe on Shuswap Lake. In the early
> season with horses, we camped in beautiful alpine meadows above 6,000
> ft [1800 m], working downhill and crawling back uphill to supper, the re-
> verse of normal procedure. Mid season, I paid off the packer and we used
> a canoe for perimeter access from Shuswap Lake....
>
> The preliminary map and photos made it possible to lay out the
> ground strips to best sample and confirm the air photo identity and loca-
> tion of various forest types and other detail, and to depart somewhat from
> the mathematical grid of random sampling. Cruise strips and traverses up
> the main valleys strengthened the control for the photo plots. Elevations
> were carried by abney level and barometer, to control quite acceptable
> contouring by stereoscopic observation of the air photos.

This summer the aerial photos provided the foundation for the forest
inventory, and the work in the field was used to corroborate the preliminary
information gathered in Victoria.

> Situated in the transition from Interior dry to wet belts, our area included
> the full range of tree species. This gave wide scope for qualitative photo
> interpretation, based on moist and dry sites and on elevation (1,145 ft
> [350 m] lake level to 7,000 ft [2100 m])....
>
> The result was some 350 square miles [900 km²] examined and
> mapped to higher accuracy and in greater detail than ever before in this
> type of survey, by a technical crew of two.

The previous summer, with a traditional large forest inventory crew, they had covered 1950 square kilometres. The 1932 survey had produced better results at a lower cost.

Andrews finished the season with a brief reconnaissance of the Bush River north of Golden.

Off to Europe

"On our return from the field, autumn 1932, Mulholland announced, regretfully, that due to impending budget cuts, we junior forest officers would be put on 'indefinite leave without pay' in the coming year." The success of the 1932 forestry survey strengthened Andrews' desire to go to Europe and learn more about forestry and photogrammetry from some of the international experts in the field. Rather than stay home, Andrews decided that he would use this leave to improve his professional expertise.

Before the 1932 field season, Andrews had written to the Imperial Forestry Institute in Oxford enquiring about post-graduate studies in air survey. In September, while he was still working in the field, Andrews wrote to Ellwood Wilson informing him of the institute's reply:

> The aerial maps have checked up remarkably well, both for location of features, and for the preliminary interpretation of forest types, age-classes, species and site. The work could never have been done in that time, so cheaply, by the old exclusive ground methods. In addition to the unique advantages characterizing aerial photographic data, a preliminary estimate indicates that field costs have been reduced by over 40% by virtue of having the aerial maps prepared in advance....
>
> Since first becoming interested in the use of aerial photographic data in our work, I have conceived the idea of seeking post-graduate training in that line. The Imperial Forestry Institute at Oxford have written that they would arrange a special six-months course for me, part of which would be spent at Oxford, and part on the Continent.

Wilson was now a forestry professor at Cornell University in New York. He was also one of the directors of Charles Lathrop Pack Forest Education Board in Washington, DC, which awarded six bursaries each year, ranging from $500 to $1500. He urged Andrews to apply, noting that several Canadians had previously received awards. Andrews applied, with F.D. Mulholland as one of his references. Unfortunately his application was not accepted. Wilson informed him that three of the six bursaries awarded were for renewals of existing grants. He urged Andrews to reapply, noting that the following year there would be one bursary for a project in aerial photography. In a letter in the fall of 1933 Wilson told Andrews that he would do everything he could to support Andrews' application. "I am very anxious to see someone put on the practical application of aerial fotografy this year."

Once Andrews knew that he was going to be put on leave, he confirmed arrangements to go to Oxford. "Anticipating unemployment, the expense of overseas study and having only recently squared off some college debts, strict economy was the rule." A relative who was in the grain-exporting business in Vancouver arranged for Andrews to work as a deckhand on a ship travelling from Victoria to London:

> With a modicum of family influence, discreet allocation of three bottles
> of Johnny Walker's Black Label (one to my Vancouver contact, one to
> the ship's agent in Victoria and one to the skipper), and a nominal fee
> [$40], I worked my passage to England via Panama on the old Norwegian
> freighter MV *George Washington*, sailing from Ogden Point, Victoria, 31
> January 1933. Mulholland had contrived to keep me on the payroll till the
> last work day prior to my departure.

The trip to England took 43 days. Andrews had not finished his 1932 government report before he left, so he worked on it during the voyage and spent his first few days in England completing it. In a letter to Mulholland in March, Andrews described the situation:

> The facilities for doing any mental or written work on the voyage were
> far from good. My work started at 5:45 am and terminated at 8:30 pm
> including Sundays. The only space available to write and spread out maps
> and summary sheets was the top of a wash stand about 14 inches square,
> or my bunk which already harboured my trunk, as well as myself. Also in
> the small cabin were an old Norwegian, and three or four of his friends
> jabbering in Scandinavian every evening. We docked at London on the
> 15th instant, and I stayed there in a hotel for four days in effort to get the
> report completed.

Then Andrews joined Professor Ray Bourne at Marlow in southern England, where some of Bourne's students were doing field studies in ecological interpretation of air photos. It took Andrews a few days to get used to the British English after hearing Norwegian spoken for six weeks. "On one occasion, when asked if I would have some 'beer', somewhat surprised, I replied that it was several years since I had eaten bear (meat). My kind interrogator looked puzzled."

He went to Oxford for a few weeks, and then Bourne sent him to the Tharandt Forest Academy near Dresden, Germany, where he had arranged a ten-week program under Professor Reinhard Hugershoff, an international expert in photogrammetry. Andrews learned about Hugershoff's work in photogrammetry and how to use some of the equipment, like the stereoplanigraph, an instrument for drawing topographic maps from aerial photographs. "Not only did the Germans have the best scientific studies but

Tharandt, southwest of Dresden, Germany.

they also had the best advanced mathematical optical studies of air photo-grammetry." Andrews spent much of his first term at Tharandt learning German. Hugershoff and a few others who spoke English assisted him. He found the equipment and studies informative, but he was mainly interested in the qualitative work related to photogrammetry, particularly for forestry.

In July Andrews re-joined Professor Bourne for a forestry tour through Switzerland. While there he spent a weekend with his sister, Mary, who was at Geneva on the staff of the Canadian Secretariat to the League of Nations. After the forestry tour Mary came to Tharandt for a week and the two of them took daily bicycle trips around the area. Andrews heard from Mul-holland that the immediate prospects for being rehired were bleak, so he returned to Tharandt, registering as a post-graduate student for the fall and winter terms.

> Mid September 1933, Hugershoff sent me up to Jena for a week-long introductory course in photogrammetry for German engineers and army officers, sponsored by the famous Zeiss firm there, the principal lecturers being Professor Otto von Gruber and himself. The array of photogram-metric equipment there included Hugershoff's Aerokartograph and an early version of the Multiplex plotter.

Zeiss produced the best equipment for transferring information from aerial photographs.

Mary Andrews, Gerry's sister, during her visit to Germany.

During the school year Andrews spent much time in Hugershoff's well-equipped photogrammetric laboratory and took several courses from the professor:

> Hugershoff's lectures were brilliant and humorous, during which he smoked cigars.... Indulgently he referred to me as his *nichtbeissenden grauer bar aus British Columbien* [the good-natured grizzly bear from British Columbia]. I became quite fond of the Professor, a strange blend of genius and fanaticism, but warm and human withal.

Hugershoff was an ardent Nazi, but Andrews did not want to get involved in the country's politics. While in Germany Andrews took his first airplane ride and commented, "I was scared stiff!" He also attended several operas in Dresden because they were inexpensive for students.

After the winter term Andrews decided to return to British Columbia. Hugershoff urged Andrews to stay one more term to complete a degree, but he was running low on money, he was starting to get homesick for British Columbia, and Mulholland had expressed optimism about his chances for re-employment.

Andrews had arrived in Germany only a few months after Hitler was elected leader. He witnessed the beginnings of Nazism in the country and was worried by what he saw. In early March he started his trip back to British Columbia by visiting Dr Christian Neumann, Hugershoff's assistant, who was temporarily seconded to the Danish government. Andrews had visited Neumann and his family several times at their home in Tharandt. "My arrival in Copenhagen by night train from Berlin, on a bright sunny morning, was with a feeling of unforgettable relief from the sustained tension and hysteria of Hitler's *nazionalsozialismus* which dominated Germany at the time." Many years later Andrews wrote in a letter: "Herr Hitler and his Brown Shirts were much in evidence and I was convinced that he was headed for war." He also recalled that, while in Germany, he had to visit the police office monthly to have his passport inspected.

From Denmark Andrews travelled to Norway to visit a friend he knew from forestry school in Toronto. Then he sailed to Scotland, where he visited the University of Edinburgh, Mulholland's alma mater, and then stayed with the parents of a friend from Vancouver. Waiting for him at their house was a letter informing him that he had received a $500 bursary from the Pack Foundation. Ellwood Wilson was to be his advisor. "Too late for an extended program in Germany, this windfall provided wider scope for professional visits en route home, in the UK, eastern USA and Canada." Professor Bourne arranged contacts for Andrews to visit in England. BC Land Surveyor R.P. Bishop had given him letters of introduction to several people in the British War Office, and he took the time to visit most of them.

When he arrived in New York on April 3, Andrews sent Wilson a telegram. While in the area he spent a day with a member of the American Fairchild Surveys Company and visited their factory. Wilson met Andrews in his home town of Philadelphia and provided him with contacts in Washington, DC. In the American capitol Andrews met Tom Gill, secretary for the Pack Foundation, along with several government officials who were prominent in air surveys for organizations like the US Forest Service and US Geological Survey. Next, he went to Ottawa and spent a week there visiting most of the notable people working with aerial photography in Canada. His one disappointment was to find that his alma mater, the University of Toronto, was not doing anything in aerial photography in their forestry program.

In a letter to the Pack Foundation Andrews wrote:

> With the background in air survey gained during the past three or four years, and that obtained from my recent trip, I feel that I am now in a position to embark on some really useful research projects in the application of the subject. My chief, Mr Mulholland, is not very optimistic about getting money from the government for such purposes, although he is personally very favourably disposed.

In June 1934, *Forestry Chronicle* (a quarterly of the Canadian Society of Forest Engineers) published Andrews' article on German applications of photogrammetry to forestry. In it he wrote: "In English speaking countries we have already accumulated a valuable store of experience in interpretation and mapping from air photographs.... The basic principles and technique also serve as a starting point for the more specialized requirements of forestry." He pointed out that German scientists used extensive mathematical and optical analysis to develop a "complicated and ingenious apparatus which produces from stereoscopic photographs, almost automatically, very detailed topographic maps on large scales with a high degree of accuracy." But he recognized that Canada was a much larger country that people wanted to map as quickly and cheaply as possible. Canadians needed to produce more portable and easy-to-use equipment that would provide accurate information of extensive areas. Much of Andrews' efforts in the coming years would be directed toward taking the information he learned in Germany and England and applying it to the forests of British Columbia.

1934

Andrews returned to Victoria on May 10, 1934, and began working for the Forest Surveys division a few days later. By this time much of the field work had been allocated. Initially he was assigned as assistant party chief in an area without aerial photographs. Then he learned about a special two-person job on a forested area with large timber trees on northern Vancouver Island. The area had been completely covered with aerial photographs and topography maps had been made from these pictures. Andrews lobbied for this assignment and Mulholland directed him to accompany forester C.D. Schultz as an air-photo specialist.

Like the 1932 field season, the preliminary work was done in Victoria. Since a topographic map had been produced from the aerial photographs, Andrews and Schultz had detailed information to use for their work. In addition, commercial timber cruises had been done in the region, providing information about the forests in the area. Before they went into the field:

> All forest types which could be safely identified by stereoscopic inspection of the photos and from a superficial knowledge of forest conditions in the Nimpkish region were plotted.... No attempt was made to distinguish timber species at this stage, although information from commercial cruises on properties throughout the area gave a good idea of what distribution might be expected. Further, only broad relative qualitative attributes were attempted.... No attempt was made to derive quantitative estimates for tree heights, stocking, etc. of stands. Most of the type boundaries were quite definite on the views so they were easy to plot accurately on the map.

Around mid July, after several weeks of office work, "Schultz and I took off in an aircraft on floats, with a new 17-foot Chestnut lashed between the floats, grub, instruments, camp kit, pack boards, and some 50 pounds [23 kg] of air photos. Major Don MacLaren was pilot and Fred Mulholland came along for the ride." The men made stops along the way to drop off caches of food at lakes they would visit during their summer's work. The first place was at Vernon Lake "where we left the canoe and a grub cache. Mulholland had to try out the canoe, which he upset for a dunking." After a stop at Woss Lake the plane flew to Schoen Lake "where Schultz and I deplaned with the balance of the outfit, on a steep salal-covered bank and wistfully watched the aircraft ease away from shore for take-off. MacLaren's farewell still rings in my ears. 'Well boys, guess you can walk home!' We did (in effect). It took more than two months." The Forest Branch report noted that: "The aeroplane in this case successfully accomplished in a few hours' work which would have otherwise required the services of a crew of packers and canoe men for two weeks or more. Using the air photos, Andrews wrote, "we could pinpoint locations precisely, choose best routes between them and best sites for volume measurements and species identity." In travelling through the rugged wet country of northern Vancouver Island, Andrews noted that, "Our real 'cross to bear' in adherence to air survey gospel was the 50 pounds of air photos, divided impartially between two packs. Weather and bush prone to being wet, a routine round our evening campfires was to spread out the photos to dry, some becoming 'toasted' to a light golden brown." Andrews also noted the contrast between his 1934 work and the previous summer. "This summer, living and travelling like two wild beasts in the coast rain forest was a distinct but refreshing contrast to my previous year in the sophisticated confines of Europe. It was gratifying, also, to have survived the vicissitudes with my partner, in harmony and mutual respect." The men finished their work in late September at Kelsey Bay.

In a letter to Ellwood Wilson in December Andrews described the 1934 field season:

> We covered 1400 sq. miles [3600 km²] in just over three months, and
> as you may imagine had to travel most of the time, the larger part of
> which was back pack over the roughest country, encumbered with heavy
> underbrush, and lacking in either trails or roads. Only the fact that we
> had a complete set of vertical photographs and a splendid topog[raphy]
> map enabled us to complete the job in the time allotted. The topog map
> which we used for a base map was prepared by the provincial Surveys
> Branch from phototopographically controlled aerial surveys. On this map
> we had plotted a preliminary forest type map from the photographs before
> going into the field. It was necessary to make only minor adjustments to
> this as a result of the field examination. We are now busy with the timber
> estimates, and management recommendations, having completed the fair
> tracings of the map just before Christmas....

In the introduction to Schultz's report Mulholland wrote: "The comparative accuracy of the report was made possible at low cost by the existence of a previously photo-topographically controlled aerial survey made by the Provincial Topographic Division. Vertical air photographs were used by the forest engineers for stereoscopic plotting of forest cover." Andrews

The famous bush pilot Don McLaren ready to leave Schoen Lake with bare-legged Fred Mulholland behind him. Mulholland had accidentally dunked himself up to his waist in Vernon Lake and removed his pants to dry. The plane, G-CASQ, was a Fokker Super Universal used by Western Canada Airways for aerial photography during the 1929 PGE Resources Survey. It operated along the BC coast until it crashed on Vancouver Island in 1938.

Andrews at Vernon Lake.

told Wilson that Mulholland could now see the value of photogrammetry
for forestry:

You will be glad to learn that at last I have been able to "put over" the
air survey concept here, morally at least. Mr Mulholland has become quite
enthusiastic, and now wants to have the whole province photographed. To
get the money to do it is of course another thing. As I see it, we must now
fight for not only more photography, but for a better kind of photography,
taken with the forestry requirements more prominently in view than has
been the case heretofore.

As his knowledge and experience in the field increased, Andrews started
to look at broader implications of aerial photographs for forestry, and at the
importance of other aspects of the equipment involved in the technology
of this new field of science. He mentioned this to Wilson:

John Liersch, the Pack Fellow from our service during 1931 and 1932,
working on selective logging on the Pacific slope, was in town for a few
days recently, and I had a very useful talk with him. I have every reason to
believe that improved air survey technique will be of tremendous value
in carrying out selective logging on the west coast. It seems to me that air
survey and selective logging must go hand in hand.

You may be interested to know that I have just completed plans for an
improved simple office stereoscope, which we hope to have made right
away. When the test model is made, and after practical tests I shall be glad
to send you further details. We estimate it will cost from $40 to $50, complete, with optically ground mirror flats.

C.D. Schultz and Gerry
Andrews at the end of
the season.

In December Andrews applied for another Pack Foundation grant to be used for: "an intensive research project here on the Pacific slope in volumetric and specific determination of forest stands from air photographs along the lines suggested in my article 'Air Survey & Forestry – Developments in Germany'." In January he received a $1000 award, with Ellwood Wilson continuing as his advisor. In his thank-you letter to the board of the Pack Foundation, Andrews wrote: "It is very gratifying to realize that the importance of air survey in forestry has received this splendid recognition. It is my sincere hope that my work under the fellowship may result in a really worthwhile contribution to our knowledge in this field."

The 1934 field season had once again proved that aerial photographs could be used to produce a more accurate forest inventory at a significantly reduced cost. Since Andrews and Schultz had better photographs and a topographic map available, the field work mainly verified the information gathered from the maps.

In his appendix to Schultz's government report, Andrews discussed the value of aerial photography:

> The large area covered, the completeness, wealth of detail, and reliability
> of the information, and the unusual quality of the maps in the Nimpkish
> Forest survey could never have been accomplished by the three-man field
> crew in three months without the aid of air photographs....
>
> In the Nimpkish project air photographs provided the basis for: (i)
> a detailed and accurate topographic base map; (ii) a remarkably good
> preliminary forest type map; (iii) efficient planning and despatch of all

field operations; (iv) accurate and universal horizontal control for locating them; and (v) a guide in appraising the reliability of (commercial) cruise information and extending it to areas for which none was available.

He noted two areas in which the photographs were deficient. "Photo-scale was too small for detail identification of species, or for data as to tree sizes." He would work at finding a solution to these problems in the next years.

Andrews concluded his account with:

> It is stated with confidence that the final maps, estimates and report of the Nimpkish Forest project, covering some 1,400 square miles [3600 km^2] of difficult, largely inaccessible country, containing one of the largest reserves of merchantable timber in the coast region, are of an order of quality comparable to the standard 1¼% forest survey on the straight terrestrial bases, at 1/6 the field cost. The maps are incomparably superior. This was in part due to available commercial cruise information, but fundamentally to the advantage of having the tract covered with aerial photographs, and to having them compiled into topographic and preliminary forest type maps prior to going into the field.

1935

In the spring of 1935 Andrews used some of his Pack Foundation money and his own vacation time to visit people involved in aerial photography in California. On Wilson's recommendation he visited Fairchild Aerial Surveys in Los Angeles, where they showed him how they were using the Hugershoff Aerokartograph and the Zeiss Stereoplanigraph for large projects like the Boulder Dam. On a boat trip to San Diego, Andrews met Jean Bergholdt, who he would marry a few years later.

In April, Wilson wrote to Andrews:

> Had a nice letter from Caverhill [chief of the BC Forest Branch] and I told him that when you got thru you would open his eyes and bring much kudos to the Department. I have been doing a lot of work on timber estimating and the more I go into it the more sure I am that it (aerial fotografy) will prove the best tool the forester ever had and it is going to revolutionize our field work and give us information soon which it would otherwise take years to get.

In July Andrews made more visits to the American northwest. When he returned he wrote a letter to Tom Gill:

> One can't help being struck by the parallelism between their problems there, and my own here, not only with respect to technical aspects, but

from the standpoint of getting a sympathetic intelligent understanding of certain administrators who have not, in all cases, been able to keep abreast of the progress and widening application of aerial photography in survey and management of forest regions.

Andrews also planned to use the Pack money to work on some of the technical aspects of aerial photogrammetry related to forestry. He wanted to take photos from several different altitudes, and use a variety of films and filters. He was particularly interested in seeing if he could accurately measure tree height and density from aerial photographs. In addition, Andrews intended to design an improved stereoscope for general mapping and forestry work.

The Fraser Mountain Lookout near Fraser Lake, under construction.

The provincial government was still feeling the effects of the Depression, and the Forest Service had limited forest inventory work. Mulholland was able to keep Andrews employed and allowed him to spend time on his aerial photography projects since there was no urgent work. Andrews used his Pack Foundation funds to cover most of the expenses related to equipment and supplies. He corresponded with several companies and government agencies regarding equipment and film for aerial photography cameras.

Andrews found an area near Victoria suitable for his project, the Sooke Lake watershed. The area was protected, had a high-quality map, and included a variety of tree types and stands. Because the area was used for drinking water, public access was restricted, but Andrews received permission to enter it to establish ground-control stations. Unfortunately, when he was ready to do the aerial photography at the end of July the company that was going to do the flying did not have camera equipment and accessories that met his specifications. He appealed to the federal government for assistance, but all of their aerial cameras were in use, and he could not get infrared aircraft film. In the end, he decided to keep the Pack money in reserve and postpone the project until the following summer.

But Andrews did enjoy some successes this season. He designed an improved stereoscope, and began working on a simple stereoplotter. "The most significant project was the determination of tree heights by simple parallax

Doug Macdougall on the road between Vanderhoof and Fort St James.

measurements on vertical air photos.... A simple parallax micrometer bar was designed and beautifully made by a local instrument machinist, Louis Omundsen." These tools made it much easier to use aerial photography to derive quantitative information about the province's forests. Andrews wrote an article about determining tree heights for *Forestry Chronicle*.

In late summer Andrews travelled with Doug Macdougall to the Vanderhoof and Fort St James area where he did some on-site photographic interpretation. After the field season he went to live with Frank Swannell and his family at their home in Victoria. "In spite of our age difference, Swannell and I shared many interests – history, language, photography and world geography." Andrews resided there for three years until his marriage.

1936

Late in 1935 the Forest Branch initiated a program of lookout photography for fire protection similar to one being implemented in the western USA. George Melrose, the new head of the Forest Protection Division, was interested in Andrews' aerial photography work. With Mulholland's approval, he asked Andrews to develop suitable equipment and procedures for lookout photography in the province.

Many of the American lookouts were accessible by road, while in BC most could only be reached by trail. Andrews modified one of the cameras used by the phototopographic survey department and placed it on the base of an old theodolite. This camera could then be levelled and turned to any angle and recorded. During the summer Andrews travelled with his assistant, Doug Macdougall, to several lookouts on Vancouver Island and the mainland.

Infrared image of Juan de Fuca Strait and the Olympic Mountains in Washington.

The work pattern was to drive by car to the foot of the lookout trail, then backpack up to the lookout for overnight. After cooking supper and camp chores, we would set up the modern theodolite and establish "true north" by observing Polaris. Next morning, we would read the true azimuths to prominent point round the horizon, for permanent reference at the station. We would then replace the theodolite with the camera, on the same tripod, orient it to cardinal directions, and expose a complete round of photos at 45 degree intervals.... Much was learned of the use of infra-red photography for recording topography from high elevation.

This was a historic year in BC, because the provincial government conducted its first aerial photography. In late September there was an urgent demand for forest cover information for a logged tract of land of about 1300 square kilometres near Nanaimo and no air photos existed.

Mulholland accepted my proposal to hire an aircraft, borrow a camera and get the pictures. We used a Waco on floats and an old World War I camera, mounted over a hole in the floor. Over a smaller hole we installed my Zeiss Ikon plate camera with ground glass for a view finder and drift indicator. We based in Victoria harbour. I acted as navigator-camera man. It took only a few days including flights to correct gaps and cloud interference. The photos were horrible, but served the purpose and proved we could get air photos when and where needed. I emphasized that with good camera equipment, we could get good photos.

In a progress report letter to the Pack Foundation, Andrews wrote:
Due to the advanced stage of the season, inadequate equipment, and lack of experience, the photographs were not all that could be desired.

Andrews setting up a camera for the lookout photography project, while Doug Macdougall looks on.

Nevertheless they met our need and the required information was obtained, fully and quickly. It created a most valuable precedent for the provincial forest service, and provided much needed practical experience, on my part, in carrying out aerial photography for forest survey.

During last two months of 1936 Andrews devoted many evening hours to designing a simple stereoplotter. He had yet to accomplish his experimental aerial photography, because he could not get suitable camera and lens equipment, but he still had almost half the stipend received in 1935.

In November, Thornton Andrews died after a lengthy illness, and Gerry returned to Winnipeg to attend his father's funeral.

1937

The events of 1937 had an important influence on Gerry Andrews' career. Through his friendship with Lorne Swannell and his family, Andrews became interested in surveying. He had used basic surveying techniques in his forestry field work and the photogrammetry he did involved mapping. The lookout photography project had similarities to the phototopographic work done by government surveyors. "Late 1936, when Frank suggested I try for a BCLS commission, I was quite amenable." Andrews passed his preliminary surveying exams and in early 1937 articled with Frank Swannell to be a surveyor. Since he was employed by the Forest Service, Andrews was unable to work on Swannell's surveying crews, but during the winter in Victoria, he spent time with Swannell learning about surveying.

In January 1937 Andrews and Swannell went to Ottawa for the annual meeting of the Canadian Society of Forest Engineers, where aerial photo-

Frank Swannell in Texas.

graphic techniques and equipment were being presented as a special feature. After the conference in Ottawa Andrews spent a week consulting with numerous air survey agencies concentrated there. Then he and Swannell travelled through the United States sightseeing and visiting some people connected to aerial photography before returning to BC.

The lookout photography project continued, and Andrews worked in the East Kootenays in the spring and early summer before turning the work over to other personnel. One of his favourite stories occurred during this time. "[Doug] Macdougall carried a neat little .22 automatic pistol with which we could knock over the odd grouse on the trail, a welcome addition to the pot for supper. When someone derisively remarked that the pistol would not be much help against a grizzly bear, Doug would reply with a poker face, 'That's just what it's for. When we meet a grizzly we can shoot ourselves with it.'"

The major development for the year was that Mulholland began implementing the province's own aerial photography program for forestry. He created an Air Survey section and placed Andrews in charge. In the spring, Mulholland gave the approval for Andrews to order from Williamson Manufacturing Company in England an Eagle III aircraft camera with full accessories for automatic operation. A forest survey on the Queen Charlotte Islands (Haida Gwaii), planned for that summer, would be facilitated by covering the area with air photographs. The camera arrived in late July. Andrews tested it, then chartered a plane for the work. Ted Dobbin, his pilot for the 1936 aerial photography, once again flew the same Waco airplane.

The Eagle III camera that Andrews ordered produced smaller negatives than the cameras generally used in North America, but the high-quality pictures could be enlarged without loss of detail. This sharp detail enabled

Fred Peterson manned Casey Lookout in the East Kootenays. Andrews recalled that Peterson wrote poetry and named his wildlife pets.

more information to be gathered from the photographs and was part of Andrews' solution to developing aerial photography in BC's mountainous terrain:

> In BC's accidented terrain oblique photo mapping, like the township system, proved inapplicable. Photo aircraft had to perform at altitudes to clear our highest mountains and detail behind these is hidden in the oblique views. Also, the perspective grid method could not tolerate distortions from high relief on the ground.
>
> The answer for BC is vertical air photography, where the camera is aimed vertically down to the ground below. This gives plan images, much like a map, with all detail clearly revealed.
>
> Another important feature of verticals is that when taken with frequency to obtain overlaps of 50 per cent plus, between successive exposures, the overlap portions may be viewed stereoscopically for an exciting 3-D effect – a perfect three-dimensional model of the terrain.
>
> This is the key for contouring and height measurements of trees for forest inventories and site classification. Today, with modern stereo-plotting instruments, both vertical and horizontal measurements of all ground features ... may be measured and mapped with remarkable accuracy.

After years of effort to get an aerial photography program started with the Forest Branch, Andrews hoped that the season would go smoothly. Instead, it was filled with many frustrations, as recorded in his diary. He spent July 8 to 13 in Victoria checking out the new camera. At the same time he was waiting for the film and filters to arrive. On the 14th Andrews left for Vancouver, where he waited for six days for the camera to be installed in the plane. On July 20 he and Ted Dobbin took a test flight, then he took a midnight boat back to Victoria to get the photos developed. Meanwhile in Prince Rupert, Hugh Hodgins could not find any 12-volt batteries needed

Andrews making a call on a forestry telephone in the East Kootenays. The lookout camera is set up down the hill from him.

to recharge some of the camera equipment, so Andrews had to send three by boat. The developed photos proved satisfactory, so he took the midnight boat back to Vancouver, and on July 22 Andrews, Ted, his wife and a few support crew headed north. After a five hour flight, with stops at Alert Bay and Bella Bella, they arrived at Prince Rupert.

Initially Andrews did a small project in the Terrace area.

> Could not locate Forestry lookout behind Terrace. Electric control for camera jammed twice on return trip over Terrace. Second time could not get it going, so operated by wind power and hand release for rest of flight. Possibly one gap on up trip at film change, and two gaps on down flight when ECB jammed.... Camera worked fine in other respects.

When they returned to Prince Rupert he spent a few hours working on the camera and eventually fixed everything.

On July 26 the crew moved to Port Clements on the Queen Charlotte Islands. Andrews optimistically wrote: "Dobbin arranged float for plane – gas in port, and all ship-shape for operations from here." There was a dance in the local church hall that night and the crew attended. The sky was clear during the night and when Andrews awoke the next morning, but fog soon rolled in and rain started falling, preventing the crew from doing any flying that day. "Dobbin advised by wire that he will be required to fly the Tweedsmuir party August 15 to 26 and they want him to freight from Bella Coola starting August 5." Governor-General Tweedsmuir was coming to visit the provincial park in central BC that was being named for him and Dobbin was the pilot for this event. Andrews would lose considerable time during the prime flying season.

A long spell of unsuitable conditions followed. Dances at night, some swimming, softball games and sightseeing occupied the time. Andrews also spent a few hours working with the camera equipment. On August

Mah Wong provided board to the crew while they stayed in Port Clements.

5 Dobbin left for Prince Rupert, expecting to return two days later. The next day Andrews wrote in his diary: "Weather suitable for photography today, especially south of Masset Inlet – but no plane." Then the weather deteriorated again.

Andrews occupied his time on the ground as best he could. One day he went to the Yakoun River to see the famous "Golden Spruce". He wrote in his diary, "Walked up trail to Locher's Farm, then rowed up the river in Locher's boat. Found tree about one mile above Locher's. Took two photos and got sample of foliage. Lowermost limbs seem to be normal with respect to colour of foliage – but upper branches definitely yellow – particularly the new shoots."

Dobbin showed up on August 14 but had to return the next day to fly in the area south of Burns Lake. The following day Andrews took a boat down to Cumshewa Inlet and joined Bill Hall on the Forest Branch's boat stationed there. On the return trip, August 24, he took the *Prince Charles* steamer to Queen Charlotte City, then went on to Port Clements.

Ted Dobbin finally arrived on August 29. The next day the weather was clear and Andrews was finally able to begin the aerial photography. "Perfect photographic day – not a cloud. Up at 5:30 – getting ready. Delayed by difficulty in gassing up but finally away about 10:30." The weather was favourable the next day, but Andrews' frustrations continued:

Another photographic day – ready to start at 7:30 am with intention of making two flights and to try to pick up Hall at Cumshewa. Plane not ready to go up till 8:30 am. No evident reason for delay. Mrs Dobbin rides with us. Very strong cross-winds from N.W. interfered with navigating our S & W strips.

Completed Strip 6 and camera jammed again on Strip 7. Found it to be the reduction gear drive, so put in the hand drives and operated the winding by hand for remainder of flight, looped around to get the gap on Strip 6. The exposure tally on the film magazine functioned improperly causing difficulty in checking the amount of film used. The electric

Allison's log dump at Cumshewa Inlet.

control exposure tally also gave trouble by jumping two or more counts on each exposure. This added to the difficulties of operation. Occasionally the EC stopped but by vigilant watching and tapping kept it going throughout.

Further trouble given by the stop watch refusing to start sometimes. Also noted that film seems to be winding too loosely on the take–up spool. Landed back at Port Clements at 12:30 noon. Dobbin says too rough to take off again. In pm started to repair camera – working till midnight.

To compound his difficulties, the strong wind caused the plane to break loose from its moorings in the late afternoon and the crew had to bring it to shore, then tie it to the breakwater.

High winds continued for the next two days, so Andrews used the time to work on the camera. During this time Dobbin left for Tlell. The wind abated on September 3, but Dobbin did not return until 4 pm. Andrews resumed the aerial photography the next day, only to discover that fog and clouds prevented him from photographing in certain areas. They flew around finding areas where the conditions were suitable.

Were up 4 hours and 25 minutes – a good flight – unfortunate that we could not complete the gap on Cumshewa peninsula.... Had no trouble with the reduction gear on the camera today. However, the counters on the magazine covers are unreliable, and the electric control gave a lot of trouble, the star wheel catch failing to engage frequently. This made it impossible to give adequate attention to the drift and interval readings, notes, and to the film changing.

Clouds and rainy weather returned and Andrews could not resume aerial photography until September 14, when they were able to get two hours in the afternoon. "The EC and reduction gears both worked without any trouble throughout but the counters on magazines and the EC box still not reliable, causing great inconvenience – had to check progress of film by taking out instrument box and reading aerial tally."

September 15 presented perfect weather for photography. The camera equipment finally functioned without any difficulty and Andrews got 4½ hours of aerial photography. The good weather continued the next day enabling another 4½ hours of work. The morning of September 17 was foggy, but it cleared and Andrews got a third day with 4½ hours of aerial photography. Everything finally seemed to be going smoothly until he received a wire from Hodgins in Victoria saying that roll 18 (some of the first aerial photographs done in late August and early September) was "a washout".

A long spell of unfavourable weather followed. On September 24 Andrews wrote: "Weather continues hopeless. Discussed with Dobbin his reaction to remaining longer in hopes of weather. He is willing to stay a reasonable while longer.... Anxious to wind up job successfully if possible." On September 29 a "violent S.E. storm broke at 4:30 am. Got up and helped Dobbin put extra mooring lines on plane, which narrowly escaped being blown against breakwater." Andrews was now ready to give up for the season. The following day he wrote: "Planned on leaving for Rupert today, but weather too unsettled to attempt crossing Strait."

[Then, October 1] dawned clear – looked good for photography, so got away, after refuelling at 11:15 – stayed up 4 hours 35 minutes. Orderly procedure of work greatly interfered with by formation of scattered cumulus clouds, and one or two strips impossible to avoid occasional tufts. Repeated Roll 18 in vicinity of Sandspit and north from Skidegate to Tlell. Camera functioned OK – but electric counter failed to turn on the even multiples of 10 – probably due to stiff lubrication – very cold at 10,000 ft [3000 m] today – thermometer between 20 and 25 [Fahrenheit; about -5°C] throughout flight.

Andrews hoped for one more day to finish all the planned work. The next day started clear but the weather closed in after a few hours. He decided to stop work for the year "due to uncertainty of weather and to exhaustion of [time] allotment – also the high risk of weather for making return flight to Vancouver." That night the community held a farewell dance for the crew that went on until after three in the morning. Andrews wrote in his diary: "Spectacular display of aurora borealis at 4 am. Weather clear."

Departing from Port Clements.

On October 3 Andrews and his crew left for Prince Rupert in the afternoon with "a friendly and warm farewell by the whole community – down on the breakwater". When they got to Prince Rupert, Dobbin became too ill to fly, so Andrews decided to leave for Vancouver on the steamer *Princess Louise*. When he reached Vancouver he found that Dobbin was still in Prince Rupert. Andrews arrived in Victoria on October 8.

Despite the difficulties, disappointments and extended stay on the Queen Charlotte Islands, the field season proved successful. The Forest Branch report stated:

> The first major aerial photographic survey was undertaken by the Forest Surveys division this year.... During the year 2700 vertical air photographs were taken and are now being plotted. Twenty-five hundred of these were taken over the portions of the Queen Charlotte Islands that had been omitted by the Royal Canadian Air Force operations in 1933. In addition, 200 photographs, covering an area of 200 square miles, were taken in connection with the Skeena River cottonwood cruise. The results of the aerial project were highly satisfactory and definitely demonstrated that the use of aerial photography in forest surveys in British Columbia can be conducted at a cost within limits that are practicable.

In a letter to Tom Gill, secretary for the Pack Foundation, Andrews described the first summer of the Forest Branch's air photography program:

> At last, after three years of effort, and disappointment, we are in possession of a fully modern outfit with lens, equipment and filters adequate to the requirements of my long cherished experiment. Our stay on the Queen Charlotte Islands was protracted into October, due to a fearful run of weather for aerial photography, but we completed our project. Operations were carried out from an altitude of 10,000 ft [3000 m], some 2000 sq. miles [5000 km²] were covered with over 2500 photographs. At the

Drying photographs.

present time, we are busily engaged in compiling the Queen Charlotte photographs, which turned out very well, and the work is progressing nicely....

The stipend of $1000 awarded in 1935 has been spun out over quite a long time, and is now practically exhausted. That money has really accomplished much more than is written against it in the financial memo attached. Confidentially, it has, time and again, served as a lever to loosen the government purse strings for equipment and operations, which without the moral force of the Pack stipend, would certainly have been turned down. The prestige which the Pack fellowship has lent to my opinions and recommendations has been no small force in promoting Air Survey progress here.

Ellwood Wilson suggested to Andrews that he apply for another Pack Foundation grant to continue his work, but Andrews declined, noting that Hugh Hodgins wished to apply for a grant.

In the fall of 1937, on the recommendation of his superiors in the Forest Branch, Andrews received two unsolicited offers to teach forestry at a university. (Several members of the Forest Branch left to teach in American and Canadian universities during the Depression years.) In one of his replies he wrote:

My efforts here during the last six years have been mainly devoted to building up a technique and staff for utilizing aerial photography in the survey, description and stock-taking of our forests. Now, after much painstaking "spade work", we are beginning to make real progress in this work. Although there is much yet to be done of a pioneering nature, the future for applied aerial photogrammetry looks very bright indeed, and the scope is almost unlimited. I am loathe to leave this work at a critical stage of its development.

The Forest Branch cabin and float plane base at Lakelse Lake, south of Terrace.

1938

Gerry Andrews' success on the Queen Charlotte Islands in 1937 enabled Mulholland to assign him full-time on aerial photography during 1938. Ted Dobbin's younger brother, Clare, was the pilot and flew an improved Waco, CF-BJR, chartered from Pacific Airways. Andrews hired Bill Hall as a third crew member, and obtained spare parts for the camera equipment.

The first project for the summer was to take aerial photographs of the area between Kitimat River and the Nass valley. Hall left a week early to make arrangements. Andrews and Dobbin departed from Vancouver on June 12, along with Mulholland. They did some photography over the Gulf Islands and stopped at Campbell River where Mulholland returned to Victoria, taking the day's photos with him.

At Prince Rupert poor weather grounded Andrews and Dobbin for a few days, but in the late afternoon of June 15 the weather cleared sufficiently to depart for Lakelse Lake south of Terrace. They met Hall at the forestry station there and began their aerial photography on the following day. Because of the high mountains they flew at 4000 metres and the temperature inside the plane was −10°C. Despite this the camera functioned successfully. A spell of poor weather followed, and Andrews probably wondered if it would keep him grounded this summer as it did last year. During this time Andrews worked up a flying schedule and did other administrative work. One day he "cut up a lot of wood." On June 20 Dobbin received an urgent telegram from Surf Inlet, a mining community on the coast. It instructed him to transport an injured man – "bleeding to death" – to a hospital. The pilot left immediately with Hall, who had been to Surf Inlet before and knew how to get there.

The Eagle III camera.

The next day was clear and Andrews was anxious to begin the aerial photography. Dobbin arrived at 11:15 and a half hour later the crew was airborne. Andrews encountered a new difficulty with taking aerial photographs in mountainous terrain at the beginning of summer. "Good light but prevalence of snow on higher ridges made light variable. Exposed for detail on timbered slopes at expense of detail in snow." Andrews also found that prolonged flying at the higher altitude was difficult without oxygen tanks. "Hall, Dobbin and self completely knocked out after landing – effect of altitude." The crew spent about a month at Lakelse taking aerial photographs that covered about 6700 square kilometres. During the latter part of June and early July they were able to fly almost every day. Then from July 4 to 14 poor weather prevented high-altitude photography. On a few of these days the crew took the local forest ranger on reconnaissance flights over his district and once they flew the local MLA, E. T. Kenney, to view the proposed road between Prince Rupert and Terrace. On July 15 the crew resumed their aerial photography. "Lots of trouble but completed operations as far as economically advisable. Due to failure of wide-angle lens and impracticability of use of the 8-inch lens on this work decided to terminate operations here in north and return south."

In mid July Dick Farrow from the Water Rights Branch joined Andrews for a few days. Mulholland had sent Andrews a telegram in late June explaining that Water Rights wanted aerial photographs of an area south of Terrace. On the crew's return flight to Vancouver, Farrow directed Andrews to take aerial photographs of Morice Lake, the Kemano River, the area around the head of Gardner Canal, and the Kimsquit. Andrews later realized that these pictures were part of the preliminary work that would become the Alcan project after World War II. Other government departments besides the Survey and Forest Branch were beginning to see the value of aerial photography. The crew stayed overnight at Alert Bay where Andrews found two RCAF aerial photography planes and crews doing photography in the

Aerial photography in northern BC in June.

vicinity. He talked to the crews about techniques and equipment they used. Andrews' crew arrived in Vancouver on July 17.

Andrews and his crew then spent about a month doing some small aerial photography projects on Vancouver Island. One important project was photographing the large Campbell River fire area so that foresters could map the extent and variable intensities of burn as well as distribution of surviving seed trees.

The season finished with about a month's work in the Okanagan valley, the first aerial photography of this area. Based at Okanagan Mission, south of Kelowna, the crew took aerial photographs from the US border up to Armstrong at the north end of the valley. Mulholland, who was in Penticton, came up to visit them before they started their work. Andrews went up with Mulholland, Lorne Swannell and a couple other foresters for a reconnaissance flight. Mulholland also accompanied the crew on their first day of aerial photography. Hugh Hodgins arrived and served as navigator for a day when Bill Hall was ill. The difficulty in doing aerial photography in the Okanagan in late summer was smoke from forest fires. George Copley and another forester "discussed possibilities of doing their grazing reconnaissance. Present conditions far too smoky to think of doing any oblique photography." After a week of poor conditions due to smoke, Andrews and the crew finally completed their work in mid September. In their book, *Three Men and a Forester*, Ken Drushka and Ian Mahood state: "The Okanagan forest survey of 1938 was the first project to use air photographic planimetry on a production scale, both in the office and in the field."

The Forest Branch report for 1938 discussed the success and expansion of its aerial survey photography program:

> This year marked a successful continuation of air survey photographic operations.... Rounding out the skeleton air survey camera equipment acquired last year, chartering of a faster, more powerful aircraft, more experienced and specialized personnel, and a season characterized by an unusual proportion of bright clear weather, all contributed to a record year for area covered. The quality of photographs and the low cost per square mile of this modern method of obtaining complete, accurate and detailed information of British Columbia's forest lands were notable....
>
> Improvements and additions to office equipment for plotting air photos have been made. After four years of development, a Forest Service stereoscope has been perfected and produced at low cost. These instruments have been supplied, upon special request, to several private and public agencies in the province and one to the surveyor general of British Honduras. An efficient and highly specialized personnel, whose value to the service increases steadily, is gradually being developed.

In summarizing the field work for the year Andrews wrote:

> The 1938 season yielded a total of some 7,100 sq. miles [18,400 km²] of photo cover including 800 sq. miles [2100 km²] of small jobs. This is nearly three times the area covered in 1937 and reflects the benefit of experience, a full season, spare equipment, an air crew of three and slightly better aircraft performance. Our accomplishment would have been better if we had done the Okanagan first where the photo season begins earlier.

During the winter of 1938, before the field season began, Gerry Andrews and Frank Swannell had taken a trip to California, where Andrews visited Jean Bergholdt again. In October, not long after the field season ended, Gerry and Jean married in White Rock, just south of Vancouver.

1939

As the possibility of war in Europe became more likely, the BC government gave top priority to locating a route for a highway through northern British Columbia to Yukon and Alaska. There had been talk of building an Alaska Highway for several years, but the Depression had made this project unfeasible. But if there was a war, the United States would want a road link to its northern territory. Surveyor general F.C. Green favoured the route through the Rocky Mountain Trench as the shortest and easiest to construct. In the summer of 1939 he hired seven survey crews to work in the area north of Finlay Forks, where the Parsnip and Finlay rivers meet to form the Peace. The Highways Department sent an engineering party. At Green's request,

Andrews' crew on the Finlay River with the Fairchild plane (left to right): Bob Fry, Bill McPhee, Mort Tier, unknown mechanic, Clare Dobbin and Bill Hall.

the chief forester arranged for Andrews and his crew to do the aerial photography from Finlay Forks to Sifton Pass, the divide between the Finlay and Kechika rivers. For this project the ministry chartered a modern Fairchild 71C, CF-BKP, specially designed for aerial photography. It also had a new two-way voice-radio system. Later, Andrews gave a mixed review of the plane: "Although the Fairchild gives good performance in the air, it is surprising poor on take-off − and this is quite a handicap on the river, which doesn't give much scope − nor margin of safety."

Clare Dobbin and Bill Hall were crew members again. Since the men were flying at 4900 metres they devised a crude oxygen supply with three tubes connected to a single bottle. "On one occasion when we began to inhale the sustaining vapours we were at once overcome with violent seizures of choking, coughing, tears and sneezing − our bottle contained ammonia, not oxygen! Immediate return to base was necessary to rectify the booboo by the supply people."

"We first covered 3300 square miles [8550 km²] from Nelson, BC, for forest inventory which allowed snow in the north to dissipate and familiarized us with our improved aircraft." This was done between July 9 and August 7. Then the crew flew to Finlay Forks, arriving there on August 9. Dick Corless, the well-known river freighter, delivered fuel and supplies to Finlay Forks and Fort Ware. Andrews and the crew had portable radio communication sets so they could get current weather reports, communicate with the other crews working in the area, and talk to officials in Prince George.

While doing aerial photography north of Fort St James, Andrews found himself flying close to the Junkers W-34, CF-ABK, flown by the famous bush pilot, Russ Baker. Nicknamed Old Faithful, this legendary bush plane flew in northern BC from 1929 to 1940.

Mulholland left the Forest Branch for private forestry work early in 1939 but Andrews kept him informed of developments during the summer. He wrote to Mulholland from Finlay Forks:

> We got only one day's photography in on this project since arriving. Nearly a week of cloudy squally weather – but the wind is swinging to the NW – so maybe we can get started again soon. I am getting a kick out of this – at one glance from the air could look down on the route we travelled by pack horse through the Pine Pass, to Fort St James – 15 years ago. Am looking forward to getting this job further ahead into new country to the North…. It is notable how badly they all want the air photographs – even the public works.

After two weeks at Finlay Forks the crew moved up to Fort Ware. Their original site was near the post. It had a short take-off distance with steep hills at the ends, so Andrews established a camp about eight kilometres downstream which had a longer channel "clear at both ends with a safe mooring for the aircraft, above which was a good campsite, fuel, water and a better view up and down the Trench. We christened it 'KP Bar' after the call letters of our aircraft." There was weekly mail service to send the film back to Victoria.

In a 1942 article, "The Alaska Highway Survey in British Columbia", Andrews described a typical day of aerial photography:

> The routine of a photographic day started with an early breakfast, a searching look at the sky, and with luck a favourable weather report by wireless from one of the survey parties up the Trench would prelude a

final review of flight plans, loading the aircraft, and the usual pre-take-off inspection. The aircraft slowly taxis upstream (wind permitting) to the bend. Then it swings around into midstream and with full throttle gains speed with the help of the current. The crew wait anxiously for the pontoons to break free from the silty flood and the sudden uplift to clear the line of tall spruce trees at the lower end of the run, sometimes by a margin none too comfortable. About halfway to working altitudes oxygen masks are put on, the hatch is opened in the floor beneath the camera, electric connections plugged in and the camera run through a few test cycles. The pilot habitually reports the whereabouts of the aircraft by wireless to Fort St James. During ascent anxious eyes scrutinize the hyperspace for cloud distribution, movement, tendency to increase or dissipate in relation to the selected area. Sometimes it is necessary to proceed with an alternative program, due to clouds interfering with the original plans.

Approaching the start, in line of the first strip, checks on drift, interval, and light are made with consequent adjustments. The photographic flight log is clocked for start and end of each strip with entries for altimeter, temperature, exposure, and haze. Intermediate entries are made according to opportunity. A synchronized watch in the camera, recording on each negative, correlates these data with the photographs. The pilot's job is to keep the aircraft on a steady course, on an even keel, and at a constant altitude. The observer, free to move about, checks the line made good by identifying landmarks in the view finder, informs the pilot of deviations from course, watches the altimeter, and sketches onto the flight map any conspicuous landmarks for controlling adjacent strips, and otherwise maintains liaison between pilot and camera operator. The latter watches exposure interval which varies with topography; head and tail winds; light meter; camera level; drift, and progress of the film in the magazine. He must be able to change magazines within the exposure interval so that the sequence of exposures in a strip is unbroken. During end turns he inspects the shutter action by removing the magazine. Another of his tasks is to keep the flight log.

The routine goes on, ploughing photographic furrows up and down the sky at 16,000 feet [4900 m], 30–40 miles [50–65 km] long and about 2 miles [3 km] apart, requiring sustained and intense concentration on the part of the crew, with only a few moments' relaxation at the end turns, until the clock warns that there is only enough fuel to make the long gliding descent back to base.

During the 1939 season Andrews' crew completed its objective, covering an approximately 25-kilometre width that included the Rocky Mountain Trench and the flanking mountains on both sides. They also did one line of oblique photographs from Sifton Pass to the Yukon boundary that

Jean and Gerry Andrews with their baby girl, Mary.

could be used for planning work for the 1940 season. "On our lonely sand bar, September 2, our tiny aircraft radio announced the ominous war news in Europe. Convinced that Hitler had 'blood in his eyes', I persuaded my crew to carry on until the season forced a halt September 24. The 1939 season proved our competence in air photo flying, if primitive by modern standards."

In his final letter to Tom Gill at the Pack Foundation in November, Andrews said:

> I wrote you just two years ago, describing our embarkation on a real air survey program. Since then we are able to report steady, if not spectacular progress.... In other words we have been justifying our existence by proving Air Survey as a real and integral part of our Forest Surveys. I no longer have to evangelize but instead have been working to capacity trying to meet the demand for air photographs and compilations of them in the form of forest maps.

Andrews now had a permanent staff of six people, with Bill Hall as his assistant. But as he finished the 1939 field season he knew that World War II would bring important changes both to his work and in his personal life. One significant event had already begun to change his life: the birth of his first daughter, Mary, in July.

Much later, Gerry Andrews summed up his forestry years in this way: "My years in forestry, 1926–1940, were good preparation for what followed, thanks to my colleagues at work, contacts with other air survey enthusiasts like Dick Farrow and to a remarkable array of frontier friends. I am profoundly grateful to Fred Mulholland, the inspiring Chief of Forest Surveys, 1925–1939."

Soldier

In the fall of 1939, Gerry Andrews had permanent employment and a baby girl, Mary, just a few months old. But his patriotism impelled him to enlist. "From my experience in Germany, 1933–34, and subsequent events I was aware that every effort must be made at once to subdue the enemy." He also firmly believed that the expertise he had developed in aerial photogrammetry in the 1930s would be useful for the Allies.

In an article called "What the War Did to Me", he described his efforts to enlist.:

> On return to civilization early October I offered my services to the RCAF.
> They declined with the excuse that the RAF was handling all air photo
> recce [reconnaissance] and intelligence. So I wrote to my friends in the
> War Office [in London] reminding them of my visit in 1934 and outlining
> my Air Survey activities since. They replied to come at once and to bring
> my assistant, Bill Hall. Red tape delayed our departure till 4 April 1940.

In a series of letters to his wife, Jean, en route to England, Andrews described the trip. "The excitement at the dock in Victoria was a bit hectic but it helped to keep a stiff upper lip. I could see you and Mary from the prow of the ship as she pulled away – till things went swimming for a moment. When I looked again we were around the corner – and away." He and Hall transfered from boat to train in Vancouver. The trip across Canada gave Andrews an opportunity to see familiar places and people.

> I woke refreshed just as dawn was breaking. We had passed Kamloops
> and were approaching Shuswap Lake. The country looked beautiful, fresh
> and still with a few pink clouds clinging to the shoulders of the higher
> mountains, promising a sunny day. This is the country I surveyed nine
> years ago, in which I used air photos for the first time – rather appropriate
> – in the string of circumstances leading to this present venture. Also it was

along the road below that my sweet bride and I travelled in that glorious October (our honeymoon 1938).

During a layover at Field he looked for friends from his two summers there, but everyone was away. "The high Rockies between there and Banff were glorious." In Banff, Andrews had a pleasant surprise. "A most charming lady was on the platform and when I stepped out she ran up and kissed me – it was Gertrude [Andrews' half-sister] – a picture of sweetness and happiness. She had her little overnight case [and] a return ticket to Calgary. We had a lovely visit and dinner in the diner – Calgary came all too soon."

In Medicine Hat he met another sister:

> I waited up till 12:30 am this morning and to my joy Leila was at Medicine Hat with Uncle Ashton, Bessie and her husband. She had got my letter just in time to make the westbound train to Medicine Hat the day before. So she and I had a grand old visit together till 3:30 am this morning – we both seemed to grab onto a mutual and mute understanding that this was a time for happiness and cheerfulness – and we laughed and chattered away – and ate a lunch of canned turkey sandwiches (her own canning).

Several more family members greeted Andrews when the train stopped at his hometown of Winnipeg: "At the station I was met by nearly all the Andrews clan that could be rounded up.... It was some confusion – fortunately the young naval lieutenant took Bill across town to this officer's mess – while the rest of us had coffee at the station restaurant. Everybody was wonderful and so kind."

And at Kenora, Ontario, "I saw Alice and Ashton Fife [more family] for a few minutes – and then went to bed. Bill says he doesn't think the CPR could get along without the support of all my relatives strung out along it."

The train trip ended at Saint John, New Brunswick, where Andrews had a brief visit with Charles Swannell, the oldest of Frank's sons. Then Andrews and Hall boarded the RMS *Duchess of Bedford*, nicknamed the "Drunken Duchess" because the ship rolled.

> We left about 2 pm, sailing straight out to sea. A destroyer escorted us till sometime during the night – and until sundown, we had a spectacular escort of three fast Hurricane fighter planes which kept zooming back and forth in perfect formation. Also there was a slower, bigger flying boat which stayed with us till almost dark. We suspect some of this "show" was for the Minister of Defence, Mr Norman Rogers, who is on board, bound for the UK. We also have two generals, two colonels, three majors, sprinkling of captains and a few lowly lieutenants. The other first-class passengers include wives going to join their soldier husbands and a handful of civilians. Second class seems to have about the same number, including a group of NCOs [non-commissioned officers] and some children.

Apparently vessels like this, which are fairly fast, do not wait for convoy.
We are over half way across now and we can put on enough speed to out-
manoeuvre a sub, should one appear. Also we carry a six-inch anti-aircraft
gun. We were entertained yesterday by a firing practice during which the
ship took the approved zigzag course – quite exciting.

By April 17 the *Duchess of Bedford* was approaching England:

We are now in the Irish Sea, having sneaked around between Ireland and
Scotland – and are heading for Liverpool – quite a narrow passage in one
place with high headlands on each side. We have passed several small ships,
trawlers or patrols. Our ship has been taking zigzag courses for two days,
a precaution against subs. We carry lifebelts at all times, even to meals. We
have also been supplied with gas masks, quite thrilling.

From Liverpool the two men travelled to London. Andrews described
how the war affected life at night:

In the blackout you see no lighted windows or store fronts, no electric
signs, no street lights. Traffic lights are little slitted +s with hoods over
them, flashing red, green. Auto lamps are a bare glimmer, just enough to
warn the approach of a vehicle. Cafes, theatres and other retreats of night
life all have a heavily curtained anteroom of darkness as you enter from
the street. Inside all is bright and gay as ever but from the street you would
think everything was shut down and deserted. It is amazing though how
smoothly the traffic is run. I suppose many people stay at home.

In May 1940 Germany invaded western Europe and its armies quickly
occupied the Netherlands, Belgium and Luxembourg. Near the end of the
month the Allies evacuated troops from Dunkirk, France to England. France
surrendered to Germany in June, and then Germany invaded the Channel
Islands between France and England. By early July Hitler began prepara-
tions for the invasion of England, known as Operation Sea Lion. Andrews'
first postings during the late spring and summer were on the southern coast
of England, an area close to the European mainland and likely to be involved in combat if Germany invaded England. During this time Andrews'
weekly letters to Jean talked mainly about the bleak military situation, but
he remained hopeful regarding the eventual outcome of the war. In early
May he wrote: "They have a balloon barrage up along this part of the coast,
and it is quite a sight to see fifty of them scattered all over the sky. Last night
we had a display of searchlights which was like a synthetic aurora borealis."
He also noted an interesting aspect of his first training centre. "A novel
feature is a women's regiment who look after the cook and other details –
their o.c. [officer in charge] is a Mrs Kirk – who is a member of, and looks
after the officers' mess – She sits with the officers – and although it seemed
strange at first, I am all for it." In another letter he mentioned his training.
"We have been very busy – the courses I am taking are excellent and will

stand me in good stead after the war is over, but they are not easy – It is a bit hard to get the old brain limbered up again to grapple with spherical trig and survey astronomy."

In mid May he commented on the beauty of the English countryside where he was stationed. "This is a most lovely part of England – to the south we can see the sea and low lying country beside – we are up quite high, and to the north is the most beautiful park-like country of green fields, little villages and tree flanked lanes and roads. All along the hill it is grazing commons with every path and road brilliantly white, because the underlying soil is chalk."

But his letter home in mid June had a much bleaker tone:

> Every time I write you, the war situation seems to look blacker and blacker in the interval – I hope, my dear that the turning point will come soon, so that each letter will mark an improvement in the situation. The loss of Paris was pretty bad – and before it happened the possibility of losing the French capital seemed terrible – However, now that it has fallen into Hitler's control, we are still fighting – we still have an organized army in France, we still have the Navy, and the Air Force, and now, every hope of reinforcements in materials from the US – in other words the fall of Paris although unpalatable, was not fatal – and we still have a chance of carrying on a longer struggle than the enemy can stand – and if all concerned keep on trying we will win eventually. By "all concerned" I mean all those who love liberty and freedom – which of course includes the US. Be of good courage my dear, and keep faith – Right will triumph!

Another letter in late June expressed similar concerns and hope.

> No doubt you have been reading the news, and follow the developments each day – and of course you will be reading that Hitler will be attacking this country in earnest soon – When this starts, you must remember that we will give him a hot reception, and you must try not to worry about me – If mail services are interrupted, my letters may be delayed, or even lost, and you will have to be prepared to endure long periods without regular news of me. It will be then that you must exercise all the faith and hope you have, because the news in the papers and on the radio may not be very good for a while – but I am sure we will survive any attack triumphantly – and eventually good news will follow, so be of good courage – all will be well. It may well be that we are approaching the turning point in the war, and that the attempt on England may prove Hitler's undoing.

After his initial training Andrews discovered, to his frustration and disappointment, that the British Royal Engineers had no particular plans to utilize his expertise in aerial photography. But Colonel Eedson Burns of the Canadian Royal Engineers heard that Andrews was in England and

Andrews with the Royal Canadian Engineers in England.

arranged for him to transfer to his unit. Although Andrews welcomed the move, it prevented him from going with Bill Hall on an overseas mission to Egypt. In a letter to Mulholland at the end of June, Andrews described his situation:

> Col. Burns was very interested in my move to the UK and rather sorry that the RE had got first choice. However, it appears now that he is arranging a transfer to get me back into the CRE on a special air-surveying job. Personally I think I can be most effective there, as survey seems to be at rather a standstill since the war has taken its recent course, and it appears that Burns has a special pigeon he wants me to help with. In the meantime I am getting some rather haphazard training – and not altogether wasting time. The move of Burns interfered with one break for me, which would have been not unwelcome, i.e. a long trip overseas – with one of the RE Survey Coys [companies] – Bill went and I was rescheduled to go up to the last minute – when Burns forestalled the thing – I was frankly disappointed, because it would have meant a most fascinating experience and some action – and a chance of promotion. However a fellow must be philosophical in the army, just as on a surveying job.

Andrews also told Mulholland that he had visited the Williamson factory a couple of times and was well received. "They remembered you – and were quite interested to learn of your experience with the Eagle cameras in BC. They have recently completed plans for a whole new line of Eagles – evidently considerably improved over the model we have – however, the outbreak of war prevented them going into production on the new series."

In early July Andrews began working with the 1st Survey Company of the Royal Canadian Engineers, although it took a while before his official transfer was completed. He wrote to Jean: "The way things have turned out I will be much better off in all respects with the Canadians, because strangely enough it has developed that they are doing some work for which my qualifications are particularly suited – and I don't think the Imperials have any special jobs of that nature – the way things have gone." His new position had another benefit. "We shall be very much better off than at present, financially." This included a marriage allowance and a monthly payment for his daughter.

Andrews was pleased with his new situation. "It has been rather nice to get into a Canadian atmosphere again, there are only seven officers in this unit – two of which went to school with me in Toronto – and among all the others I have many mutual acquaintances – so that it is very congenial. They are an exceptionally fine bunch of men, as of course field engineers usually are." But it took time for him to begin doing any work related to his experience, and he expressed his frustration to his wife:

> It seems stupid that so far I have really been doing odd jobs requiring no
> special skill – it has been interesting, and good experience, but I feel the
> time has come when they should assign me to air survey work, in which
> I have so much unique experience and training. Things seem to take so
> long in the army. The God Red Tape moves in many mysterious ways, and
> so many authorities must have a finger in each pie. In my case, too many
> cooks are threatening to spoil the broth.

In August an old friend arrived in England. "Lorne Swannell called me on the telephone the other day – he is located not too far from us, and we are going to try to get together for a weekend. He had a good crossing – it's just possible I may be going down to where he is on business tomorrow. It would be great if we could arrange a leave together. Lorne would be ideal company." In mid August Andrews went to London for three days. Swannell joined him for part of the visit and they went to a play together. Andrews was in London when the first bombing occurred near the city:

> As you will know from the papers and radio the outskirts of London
> had had their first taste of bombing – actually the damage was not great
> – but it was very reassuring to note how calm and cool the London pop-
> ulace is, now that danger has been brought to their door step so to speak
> – I was there during two air raid alarms the previous day – of course
> nothing happened, but the people seemed to take it as a matter of course.
> During the last few days I have felt more convinced than ever that Hitler
> will never gain a foothold on this island – and he will never subdue these
> splendid people. Of course the times are tense and anxious – but we must
> expect that.

In late August, Andrews learned that Swannell's unit was stationed nearby:

> Yesterday Lorne arrived at his unit, which is only a mile away from us – so we are practically neighbours – that is a bit of alright! We are anticipating spending a lot of our spare time together. Today we went to see a large estate nearby which was open to the public for one shilling a piece in aid of some war fund. It was lovely.... We also found some excellent blackberries which we took advantage of.

Lorne Swannell.

In a letter to Mulholland in late August, Andrews wrote about his feelings for British Columbia:

> Your letter of 28 July was most welcome, and written as it was from Skaha Lake – out there in our good old BC hills – the things you say – and all – well in current phraseology, it made a "direct hit".... One thing, it does a fellow no harm to get a distance back from it both in time and miles and circumstances too – as I find myself now – and take a good contemplative perspective view of it. What a country – and something to come back to – regardless of the human superstructure – for its own sake – the hills, lakes, streams, forests – and the loons and the thunderstorms swooping around old Baldy's shoulder – you bet! Each time I have left BC on a temporary expedition it has pulled harder than ever – and I doubt if I can resist it any more this time than after the Toronto years or the German interlude.

He was pleased to learn that Jack Burton had continued doing aerial photography in BC that year:

> I am glad that work is being carried on – it all helps to keep our Air Survey baby alive and nourished – God knows what it will grow into once this business with Hitler is disposed of. Every time I see our bombers and fighters buzzing around overhead like a swarm of hornets, I can't help thinking what we could do with a couple of them on air photography over BC.

And he expressed to Mulholland his frustration at the slow pace of development:

> I have had rather a trying time getting straightened out to put my full energy, training, qualifications and interest into this fight – in air photo

Andrews in a reflective moment.

interpretation and intelligence – partly through what appears to be a red tape impasse – and other unfortunate circumstances. Particularly aggravating is the knowledge that air photo interpretation is of greatest importance – and those who are trying to do something with it are desperately short of experienced and qualified personnel.

As the German bombing of England increased during the late summer of 1940, Andrews wrote to Jean:

As you know from the radio and the papers bombing has been intensified recently – and it is a beastly business – but the more they bomb, the more determined our people seem to be to keep up their courage and determination to tough it out and eventually strike back at Hitler. This aspect of modern warfare appears to me as most despicable and rotten, when innocent civilians – children, women, old men and young boys – factory workers – are exposed to attack as mercilessly as the fighting forces. What terrible weapons "civilization" has brought us.... How monstrous is modern war which makes the sky – one of the most beautiful and heavenly features of our world – into a source of death and destruction. I sometimes look up into it at night, when the drone of German bombers can be heard – see the search lights wan probing fingers converging on a spot – and beyond see the stars of the firmament – still, cold, unchanging in their age-old arrangements looking down as it were ironically. At times like this it makes you wonder what stupid things men are in their ghastly doing – what futility.

In late October Andrews had a pleasant diversion from his work:

I have ... been given permission to take a week's leave starting next Tuesday and Lorne has been able to get his to coincide – so we are going together to Scotland and back. This will take a bit of cash even though we do get our railway fares paid.... Today Lorne and I went for a good tramp before lunch – it was cool and dry – just perfect for a good brisk walk – did about five miles.

The two men had a wonderful week visiting Scotland, but when he returned Andrews found that his war-time work was still frustrating:

There is some talk of having me placed in some work of my own line – but just talk. Boy! Things move slowly in the army – if the war lasts long enough I may eventually be allowed to do the kind of work which I came over to do – it is very trying on one's patience....The Canadian people hauled me out of the REs and spoiled a very interesting program for me in Egypt and since have made absolutely no use of the special qualifications I have, and for which they went to the trouble of having me transferred – looks to me like a lot of stupidity and confusion. Certainly my experience with my own countryman's army so far has been far from inspiring. I have had to use an awful lot of philosophy, patience and even cynicism to keep up faith in my particular case.... However, I am stubborn enough to keep on trying to follow up my own air survey line even if it means sacrificing opportunities for advancement. I'll get there yet dear – damn them all, but in the meantime I'm getting hardly fit to live with. You should think me an awful wretch if you really knew how mean and cranky I have become – a regular bear. Lucky you are so far away – you would be quite disgusted.

I had a letter from Bill Hall who is very happy and interested in his work, and he said that they were very disappointed that I had not arrived with him! Well I'm glad Bill got a break and has escaped the unending delays that have been my experience with the Canadian set-up.

Andrews was able to get leave for a few days at Christmas. He described his return trip to Jean:

We didn't get back till late last night – travelling nearly half of the 215 miles in the black-out, through the maze of twisty roads and villages. It is really remarkable how good one gets with a little practice in black-out driving. As we came down west of London, we could see the ruddy glow of fires in the capital caused by an unusually heavy blitz. There is something sinister and diabolical in that sight.

In a letter to Frank Swannell he wrote about evidence of ancient surveying in England. "We travelled 80 miles [130 km] along Watling Street, the old Roman road from London to Chester. It was quite a thrill following long tangents for miles at a time. Presumably Caesar was not troubled with property rights when he laid out the road."

In early 1941 Andrews received a new position and finally started doing work related to aerial photography. He presented his ideas to the Canadian General Andrew McNaughton for "improving optics etc. for military air photo intelligence". McNaughton had been head of the National Research

Council from 1935 to 1939 and had made advances in the science of artillery during World War I. He had an understanding and appreciation of Andrews' ideas. With the general's authorization Andrews would

> consult with the Ross Optical Co., Kodak Research, Adam Hilger and
> the Williamson Mfg. Co. in London and Chance Bros Glass Works in
> Birmingham, with a view to producing improved small-size air cameras,
> especially for the Canadian Army O/S. In these contacts I met some won-
> derful scientists. My philosophy, which the General shared, was that a small
> camera giving high resolution 5x5 negs, implied less than ¼ the weight
> and bulk of the orthodox 9x9 camera. The smaller negs could be projec-
> tion printed to the standard 9x9 format. This would allow multi-camera
> installations in aircraft. I had applied the small camera principle success-
> fully in BC for three seasons, 1937, 38 and 39.

In February Andrews wrote to Jean about his new position.

> Have been just over a month now on my new job and been appointed
> officially on the "I" [Intelligence] staff.... One thing about the job, I have
> been busy all the time, and have been able to get around to a certain
> extent to see how things are being done by the Canadians and British Air
> Force in the way of air photography, so that is something.

He noted that Lyle Trorey, who had worked with him in the Forest Branch, was the new person in his position. He also met another person from BC:

> On Sunday I went up to Lorne's battery with Capt. Rothery and was
> pleasantly surprised to find Dick Farrow there – he belongs to the same
> regiment. So we had quite a reunion and stayed to lunch. Lorne gave me
> a book which Frank Swannell had sent for my birthday – some very use-
> ful math tables – so when you see Frank tell him I have it and am most
> pleased – and will write him in due course.

In another letter to his wife Andrews expressed more admiration for Londoners:

> When we left London tonight some of the people were beginning to take
> up their places in the underground for the night. My I do feel so sorry
> for these poor folk and I certainly think they should be compelled to
> vacate all children from there. The funny thing is that the Cockneys just
> don't seem to be able to bring themselves to leaving London – it is their
> universe – and it would be just the same as if you and I were ordered to
> send Mary to some other planet! Another thing too – they are tough!
> The survival of the fittest through so many generations in London's dust,
> smoke and fog seems to have immunized them from catching microbes in
> the foul underground air – am sure it would kill me in no time!

His letters to Jean during the winter and spring of 1941 became more optimistic about his work. In one of them he said:

I am trying to have some reforms put over in our air photo techniques
etc. – and so far the old man seems to think my ideas are not too bad. If
I can be allowed to work away on things of this nature and get my work
laid before the right authorities – well then I feel that I am doing something worthwhile – regardless of what rank and pay the "office boys" see
fit to give me. At last I feel as though I am getting my ear in – I hope it's
not too late.

Later he wrote, "Another busy week, and a useful one. If all this work
comes to a fruitful result, then it will be worthwhile and I will feel that I
have made a small contribution to the cause. If not, then at least I will have
had some good experience out of it, which will be useful in the future."

And on April 20 he talked about his first year in England:

How happy we will be when it's all over. On the other hand, I should not
want to come home now, with the job unfinished. Things are a bit tough
just now for the British and their game but small allies. It is a big job – a
great deal bigger than many people realize – to bring Hitler and his gang
to a standstill and drive them back into their own country – and out. We
are feeling the impact of his desperate effort to save the collapse that he
must know is imminent if he doesn't force a decision this year.

But his work kept him occupied. "The weeks seem to click by at a terrific
rate."

Lorne Swannell was now stationed farther away, but Andrews still visited
him occasionally. "Got in touch with Lorne yesterday by telephone, and arranged to go over by motor bike this morning to see him. Dick Farrow was
there, and we had an exchange of notes. Andrews also received letters from
Bill Hall in North Africa and, despite his recent optimism regarding his own
work, he expressed some regrets in a letter home:

When I hear from Bill, it makes me rather dissatisfied with my army history since he left. As he says, we never should have split up. However ...
my time since New Years has been very worthwhile, and in the long run,
more important than if I had gone to Africa. Also, we have been much
better off financially, which is an important consideration for a married
man. Still I envy Bill, seeing a new part of the world, and taking part in an
active and fascinating campaign.

In August 1941 Andrews was promoted to captain, assistant to the director of surveys for the 1st Canadian Corps under Lieutenant Colonel Jerry
Meuser. He also had the Forest Branch send over one of his stereoscopes. "It
is better than anything we have seen here, and should cause some interest.
The Lands Dept. seem keen to help out in this way, and it sure gives me a
fine feeling to know that they are backing you up in what we have to do
over here." (Periodically the Forest Branch sent Andrews parcels containing
cigarettes or candy, and they also paid Andrews' superannuation.)

Despite his promotion Andrews still felt some frustration in advancing his advocacy for aerial photography. In a letter to Mulholland in late September he detailed his difficulties:

> So far, getting survey streamlined in the Canadian Army has been uphill business. The trouble is partly that for training and operations here in England the maps are too good, and nobody is forced to think what it will be like when we get into country where maps are inadequate. As you know, there is probably no country in the world as well mapped as G. Britain. So when you try to argue for better surveying cameras, aircraft and personnel organization to take air survey photographs the reply is always – "What do we want survey for anyway?" More than ever do I realize what a tremendous advantage it was to be part of a small flexible organization when, with your backing, we organized our Air Survey Section in BC. To get a recommendation put through into action in this show requires so much convincing and liaison and co-ordination. At any rate we do our best. A single, well informed authority controlling all the phases of air survey from ground control to printing – including the flying, the camera equipment, aircraft and installations is what is required.
>
> There have been several fundamental advances in technical equipment for air survey – but it seems impossible to utilize what the scientists and technicians have to offer us. Another difficulty is that air survey involves two services, the Army Sappers and the Air Force. The result is fatal. However, I'm stubborn as ever, and keep trying – already have got some advances put into effect, minor details but all part of the main scheme, so that if and when we get the right organization, we will be in a position to go right ahead.
>
> To compensate many discouragements, I have had the pleasure of liaising with many interesting people.... In many ways it has been like a unique post graduate course, much of it I owe to General McNaughton who has put me into a position where I could do this sort of thing since last January.
>
> Well F.D. often wish I could have a chat with you – God knows when this war will end. From our point of view it seems to be just getting started. 1941 has been better than 1940 though!

Mulholland replied in November with encouragements, but also a prescient view of what would happen after the war:

> I am extremely pleased you are doing something that is of real value and that will keep your brains and organizing ability in practice; you will need them when you get back; as to that, I think you and the other warriors are going to meet very much the same conditions as existed after the last war and you will have to stick together and demand your rights. It looks as if a number of fellows, and not all old staggers like me, either,

are determined to stay at home and make hay while the sun shines and I find it hard to believe that they will be any more ready to give way for the returning soldiers than their prototypes were in 1919.

While Andrews was overseas, Jean maintained the household in Victoria. Since she had only moved from California in 1938 she did not know many people in the city. She also had not worked before. But the family needed to supplement Gerry's army income, particularly when his salary was delayed for a few months during his transfer from the British army to the Canadian. Andrews' friends helped Jean get work in the mail room of the BC government and arranged for the sale of the Andrews' car. Jean used the money to purchase a lot, and the BC government helped her gain access to an assistance program for military families so that she could build a house on the property. The Garman and Phipps families provided accommodation for a while (Eric Garman was a researcher in the Forest Branch and Al Phipps worked on Frank Swannell's surveying crew), and the Swannell family often invited Jean for meals and company.

The bombing of Pearl Harbour and the declaration of war between Canada and Japan dramatically changed events for the Andrews' family in Victoria. The house on Marlborough Street was only a block from the ocean and suddenly appeared vulnerable. Gerry worried about his wife and daughter, but also expressed confidence to Jean in his letter of December 14, 1941:

> After the kite went up in the Pacific a week ago I thought several times of cabling you and held back waiting for more news and the picture to clarify. What irony – that my loved ones are now right on the front – and here we seem to be safer than anywhere in this world. There is probably no place on this globe so formidably defended as this country right now. What a week it has been – I refused to let myself get in a panic – and now I feel calm and confident. I think seriously that there will be no attempt at the west coast of the US or Canada. On top of that I feel every confidence that the defence jointly between the States and Canada will be more than adequate. Japan will be at a great disadvantage in attempting such a remote control attack – and no doubt she's wise enough to concentrate her strength in what she wants to make her own sphere – the western Pacific. If however at any time a surprise attack should be attempted on Victoria or the vicinity please cable me how you are – because it will be a long anxious wait for a letter to catch up with the news of anything like that.

> No, the thing that bothers me is that you will have to suffer the misery of the black-out. It is miserable too – especially the first winter, and especially in your case, with Christmas coming along – when lights are such an important feature of Christmas cheer and happiness. But the worst thing about the black-out is the danger in the streets at night. You

will be very careful I know. Always carry a small torch in your purse – even in daylight – because one invariably gets delayed and has to come home in the dark. When you cross the road or when you even step off the sidewalk, shine the light on your own feet – it helps the motorist see what you are and where you are – and of course never wave it around pointing the beacon into the face of oncoming vehicles. Bicycles are as dangerous as cars too – even more so, because they don't make any noise, and they often have bad lighting.

The black-out will put a cramp in your social life and you will likely tend to spend more time at home in the evenings. Perhaps you had better get the gramophone after all.

I don't mind the black-out nearly as much this year as last – but it's gloomy business at that.

In March 1942 Andrews gave a lecture to the Royal Geographical Society on aerial photography on the Alaska Highway route. He also wrote an article on the topic for a British magazine that published it in July. That same month he received another promotion:

As you see, I have been made an Acting Major, if I am a good boy and keep my present job for a few months, it may be confirmed as a perma-nent rank for the duration. It came out in orders last week, Friday, and I wasn't expecting it so soon.... My official title now is DAD Svy which means "Deputy Assistant Director of Survey" for the First Canadian Army....

I didn't get nearly as much kick out of this promotion as I did when made a captain. I think a captain is a nice rank, it kind of distinguishes you from the common run, and yet you are still one of the boys. A major is supposed to be just a bit more dignified. However, I'll try to live it down, and hope I will be able to justify the promotion.

In a letter to Jean in September Andrews wrote about a connection be-tween his aerial photography in BC and the war. "One interesting thing, the Sitka spruce that I surveyed from the air in the Queen Charlotte Islands in '37, has now come across my path in a very important way, and the experts whom I have been seeing in the manufacturing end were very interested to hear about the story behind the important wood they are using to help beat the Nazis." (Spruce wood was being used to build Mosquito bombers.)

During the fall of 1942 Andrews achieved a major success in his quest to develop the aerial photography work in England. "With General MacNaughton's support I found time to develop specs for 25 Eagle V cam-era units to be made especially for the Canadian Army O/S by Williamson's. These included several optical innovations as well as a between-the-lens shutter." Andrews conveyed his excitement to Jean in a letter in early No-vember. "It looks as though I've won a battle I have been fighting practically

Airgraphs

In mid December 1941 Andrews received four airgraphs from Victoria. The airgraph was a new technology where a letter was typed or written on a special one-page airgraph sheet with postage placed on the back of the form. The airgraph was taken to a processing station where it was photographed and placed on a film reel that could hold over 1500 letters and weighed less than 200 grams. These reels were sent by air to Great Britain where they were printed and mailed. This technology significantly reduced the time it took for mail to travel between Canada and Great Britain.

Andrews commented on the service in letters to Jean:

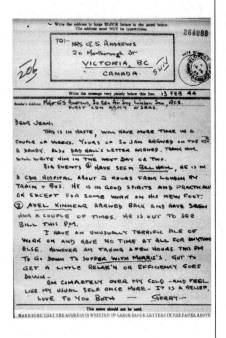

> The airgraphs are excellent, fully legible – and a lot on each one, especially if they are type-written.... [They] have the great advantage of taking only half the time to come, and that makes you seem just that much closer. They are a little brief, but you get a great deal of news in the space allotted. I think they are best for the regularly weekly letter. If you happen to be in a communicative frame of mind, and have the time, at the time, a longer detailed letter is nice to get too, occasionally, as a special treat, but for the regular job, I think the airgraphs are the thing.

Initially the service was only available from Canada to Great Britain. But in July 1942 Andrews wrote, "There seems to be some talk of inaugurating a westbound air letter service, and if they do, we will be at least five weeks closer to each other." He sent his first airgraph to Jean in early August.

ever since getting over here.... This time it really looks as though the thing
is put over.... Someday I'll be able to tell you the whole story, but for now, I
know you will be glad to share my success." The following week he wrote:

> Mr Colin Williamson, head of the firm who make the Eagle air cameras
> phoned up on Friday inviting me to come up to his home at Bourne
> End, a beautiful part of the upper Thames valley.... He thought it would
> be a good chance to talk over more leisurely some rather important and
> intricate business....
>
> Had to go back to town this pm to the meeting of the Royal Geo-
> graphical Society to hear Dick Farrow give his paper on snow surveys
> in BC.... He and I had dinner together afterwards, and then I caught the
> train back here, as I wanted to get this letter off before too late. It was
> a lovely crisp fall day again, and tonight the stars are in full force over-
> head. The shortage of petrol has greatly reduced the traffic on the roads,
> especially after dark, which makes it much safer getting about the streets
> in the blackout.... The darkness has probably for the first time in their lives
> caused city folk to notice and appreciate the grandeur of the heavens on a
> clear dark night.

In late November Andrews had a special occasion:

> Today has been a gala day. Axel [Kinnear, a BC Forest Branch employee
> who worked with Andrews in 1939] came up for the day. He is a full-
> fledged Lieut. now, and I feel that the Cdn Army has gained a fine officer
> in him. Dick and Lorne phoned that they would be in the district for
> pm tea, and would meet me at Morris'. It turned out to be practically
> an invasion as the tea party included Axel, Lorne, Dick, Bert Hammond,
> myself, and Col Carrie, so to help out with this sudden emergency, I took
> up your cake, and it saved the day royally. Mrs Morris was in high fettle,
> as well as Ecila, and everybody had a good time. I think it was a record to
> have 5 Victorians together at Morris' at one time.

From late fall 1942 through the spring of 1943 Andrews consulted "with
the Ross Optical Co., Adam Hilger, Eastman Kodak and Williamson Mfg
Co. in London, Chance Bros. Glass Works in Birmingham and the RAF
Research Establishment, Farnborough, re improved optics, mechanics and
photo chemistry for a super small size air camera." The Eagle V camera
model would incorporate the most modern technology available to produce high quality photographs. The smaller size would enable three cameras to be installed in a plane. Andrews wrote to his wife: "I continue to be
busy ... fighting a battle all my own these days – not a very important one
perhaps, when I look around and see what really big things are taking place
in various parts of the world, it makes my business look like a storm in a
teacup. However, if I can do my small part, and others do theirs, small and
big, we'll win, and all the sooner." In another letter home he noted recent

US victories against Japan: "It relieves me of a lot of anxiety about my wife and baby and our little house on the Pacific coast."

In June 1943 General McNaughton assigned Andrews to mobilize and command the 10th Canadian Air Survey Liaison Section, to finalize the Canadian Army's order of 25 Eagle V Air cameras, and to monitor their production. "This allowed giving all my time to the camera project with technical help whom I had to train." Andrews' section only had about 12 people and was "attached to larger units for messing and accommodation. We had several moves, but all were handy for contacts in London." He explained to Jean: "It really means a continuation of the work I've been doing all along, but this more or less makes it official.... It will be interesting to get a brand new thing organized and run it the way that seems right. It is a very small and modest little unit, but a very specialized and important one."

In August the general sent Andrews to Ottawa to consult with National Research Council scientists regarding a tri-camera mount for the new camera. Unlike 1940, he flew across the Atlantic Ocean in a four engine plane from London to Montreal. While he was in Canada he was able to travel to Victoria and spend about two weeks with his family. When he returned to England, also by air, one of his first activities was to present a BBC broadcast for school children about the Alaska Highway.

In November 1943 Andrews' section moved to London with all personnel billeted in several homes. The Eagle V cameras were now in production and the men monitored and checked the equipment. They also worked on developing a better stereoplotter – Andrews had brought his back with him from Victoria. In December General McNaughton, who had been a strong advocate for Andrews' work, was recalled to Ottawa, but by then the production of the Eagle V cameras was almost finished. On New Year's Day 1944 Andrews received the MBE (Member of British Empire) award. He wrote to his wife that the award surprised him and that it was probably "an acknowledgement of plugging away at the job, being a bit obstinate (in my case)".

Later in January Andrews' section was changed to 30 Canadian Air Survey Liaison Section to avoid confusion with a similar Royal Engineers unit and was transferred to the Canadian Military Headquarters. Andrews' expertise in aerial photography was now going to be used by the Allies. As the Allied commanders began planning Operation Overlord, the invasion of continental Europe, they realized that they did not have adequate maps of the beaches along the coast of France, and would not be able to obtain them by conventional methods. Later Andrews reflected on the situation:

> Once the concept of OVERLORD crystallized, landing massive assault forces on the Normandy Beaches, there was a dire need for detailed, reliable information about the submerged approaches thereto. How far

out from "dry" land would landing craft ground, according to draft and a given tide level? After discharging their precious cargoes of troops and equipment, how far and how deep would the wading be to dry land – under lethal fire? Would they start waist-deep and then encounter depths perhaps overhead? Marine charts of the English Channel, adequate for merchant shipping, lacked the required littoral detail close to shore. Meagre detail charts dated mostly from Napoleon's time, since which, for over a century, the Channel had been swept by countless storms and tides – Nature's bulldozers of friable ground.

All attempts to obtain detail sounding were subject to formidable enemy occupation. Submarines under cover of darkness or fog were inadequate for soundings. However, air photography, by stealth and surprise, offered possibilities....

As early as 1942, Major WW Williams of Cambridge University proposed the use of waves approaching the beach from deep water, the velocities of which decelerate and the distance between them decrease with the shallowing depths, according to a classic hydrodynamic equation....

The velocity of the waves approaching the beach could be used to determine the depth of water below. Since the velocity could be measured on timed overlapping air photographs, the hydrodynamic equation and photos could be combined to make a map of the beach.

Initially aerial photographs were made along the coast of southwestern England:

Field tests at Bideford Bay, Devon, in 1943 confirmed Major Williams' proposals....The image of each wave appeared in at least three successive timed photos. From decreasing velocities and wave lengths corresponding depths were derived to give a profile of the bottom, from the outermost waves to the beach. This was correlated to chart datum for known tide level at the time of photography. Results were checked against actual echo-soundings along the same line. Accuracy of about one foot in ten to five feet in twenty-five were indicated, and considered accurate. Those concerned at Supreme Headquarters Allied Expeditionary Force (SHAEF) at Bushy Park, near London were satisfied, and the go ahead was given for operational application of Major Williams' "wave method".

In January 1944 Andrews' section was assigned the task of mapping beaches along the Normandy coast. When the Allies began taking aerial photographs of the French coast in February Andrews was disappointed to find that obsolete American cameras were used to take the aerial photographs instead of the new Eagle V cameras.

When General McNaughton relinquished command of the Cdn Army Overseas, late December 1943 ... our Canadian air survey developments

One of the aerial
photographs used to
map the Normandy
beaches, considered a
good example of well-
defined wave action.
(From "Memoran-
dum on Photography
for Beach Gradient
Dertmination".)

lost an influential champion. Otherwise the specifications and planning
for wave velocity photography might have utilized the new Canadian
equipment and know-how to advantage. Instead, surplus and obsolescent
American cameras were modified and used.

The aerial photographs were taken from top-performance Spitfire airplanes. The photo run was along the beach to include about half land and
half ocean. Andrews described the procedure:

To avoid detection, approach to target area was just below "con-
trail" height, about 30,000 ft [9100 m]. After alignment with the target
beach, the pilot power dives to about 14,000 feet [4200 m], levels off,
and switches the camera on. Air speed along the photo run decelerates
from about 380 mph to 320 [610 to 515 kph], the camera automatically
exposing a new frame every four seconds while the pilot endeavours to
hold course. The run is completed in just a few minutes before the Enemy
realizes what's up, and the aircraft peels off to seaward, drops to minimum
height, and heads back to Benson, like a homing pigeon. Between 5 Feb
and 25 April 1944, 12 sorties were made, happily without loss. Only 5 sorties yielded usable photos.

Andrews' section worked on the aerial photographs in secrecy. He told
his men that their work might only be a diversion by the Allied command
and that they could not be sure the information from their project would be
used. Even Andrews did not know for certain, although he was reasonably

confident that the Allies needed these maps for the planned invasion of France. He observed: "Liaison with SHAEF at Bushy Park was by Lt Colonel W.E. Browne, MBE, a fine New Zealander. The avidity with which he whisked our profiles to SHAEF suggested the importance of our work."

In March the unit moved from London to a large secluded country house where the men averaged 12-hour working days. To keep Andrews' section as anonymous and low profile as possible there were no unit signs posted nor any support troops provided. The men did their own housekeeping. Andrews commented to Jean: "We are like one family now, quite self-contained and independent. If I could only get a cook like the old boys we used to get for our survey parties out in BC it would be ideal. I realize since being in the army, what wonderful men those old survey cooks were." In a later article he said about his men:

> all took pride in the work as vital and appreciated the advantages we
> enjoyed – minimum "regimental" routine, informality and pride in the
> unique techniques conceived and applied.... Liaison with SHAEF was
> agreeably maintained by Lt Col William E. Browne ... who would visit
> once or twice a week to review progress and problems, and voraciously
> gather up the current batch of finished profiles.

When they began using the aerial photographs to map the Normandy coast they found that tilt of the airplane could distort the photo scale, and they needed to develop a procedure to correct this problem:

> Another refinement we called "the mean wave adjustment" to remedy
> minor photo-scale errors. Finally, on the assumption that the period for a
> system of waves from a particular sortie was constant, a routine of har-
> monizing observed wave lengths and velocities to conform to an overall
> mean period was applied. Thus, the vagaries and irregularities of wave
> incidence in Nature as described by the Psalmist, were disciplined to serve
> more closely the mathematical relationship of wave length, velocity and
> depth. All these refinements were progressively developed and applied.

Although Andrews couldn't tell Jean about the work his section was doing he did describe the conditions and tension:

> It's work, work, work, as it should be, but there isn't much to write about.
> It's going better all the time, and even more interesting too, because we
> are learning more about it all the time. Of course there are the bad mo-
> ments, when nobody seems to be able to do a thing right, even myself.... I
> make it a point of letting my boys have the Sunday off, they need one day
> a week off the work, and the work is the kind that suffers quickly if the
> crew get fagged. It calls for great concentration and care, and their wits
> must be bright and sharp or the work suffers.

In another letter he wrote: "The work has been going fine, and results seem to be improving all the time. The boys take a lot of interest in it too,

Andrews (front, centre) with his section. Bill Hall sits next to him on the left and Howard Davis, from Victoria, stands behind at the far left. Duff Wight, who worked for Andrews after the war, sits in the second row on the far right.

especially as we have developed the technique ourselves which has given the best results. They are always rushing us tho' so maybe it is of some use, which is gratifying too."

Andrews, too, needed time to relax: "If I don't take advantage of Sunday afternoon and evening to get out, the work suffers." In his free time he took walks, bicycle rides or did physical work to relieve the tension. In February his close friend, Bill Hall, who had been wounded and lost a foot in the Italian campaign, was transferred to a hospital near London for rehabilitation. Andrews went to visit him as often as possible. In July, when Hall was able to resume working, Andrews attached him to his section.

When the Allies landed at Normandy on D-Day Andrews learned that his section's work had been used:

> None of the Unit will forget the night 5/6 June 1944.... Next morning, when the 8 o'clock BBC broadcast announced the Allied assault on the Normandy beaches had begun an hour earlier, the responsive roar from my boys, in unison, was "OUR BEACHES!" It was a climactic moment. As later broadcasts ensued, our jubilation was tempered by grim appreciation of Enemy resistance, and concern for the troops involved. We hoped our profiles might have helped to lessen the deadly toll there.

He described this historic event in a letter written to his wife on the same day:

> You can imagine the excitement here, the sergeant came into my office this morning just after breakfast and said that the German radio was announcing the landing of Allied troops on the French coast. Later

followed the official announcement from Eisenhower's HQ, and then the usual news bulletins. My draftsman who is pinch-hitting as cook for us has the only radio in the establishment, so we congregated in the kitchen for the various broadcasts. They all say there was a terrific amount of air activity during the night but afraid I slept through it all. However, today, the skies overhead have thundered with recurrent waves of aircraft. Of course we hear and see only those which happen to have their course near our little spot of England. It must be terrific. I couldn't help thinking of the lads making their way across the beaches of the French coast, they are probably doing it this minute too, some will fall.... They will go on and on and when they have gone enough, it will be the beginning of the end. And that's what the whole world is waiting for.

You may be thinking that we who are here, not yet in it feel out of the picture, but in my little section we are in very high spirits and someday I'll be able to tell you why. One of the boys has his birthday today, and his name begins with D. Your cake arrived yesterday, so we had some of it for supper tonight by way of celebration.

In their book *Three Men and a Forester*, Ian Mahood and Ken Drushka note the significance of personnel from British Columbia in the war effort:

> While air photography and mapping were pre-war intelligence tools, the science was underdeveloped and not adaptable to mass production technology under battlefield conditions. It was British Columbians who pioneered and expanded military technology for use in sight and sound of the guns. It was a British Columbian who led the First Canadian Air Survey Company. That we could establish the objective of "a map a day the day needed" and make it work was a singular achievement, much of it shaped by the forest survey technology pioneered in BC.

The work of Andrews and his section is credited with playing an important role in the planning of the Allied D-Day landings and saving the lives of many soldiers.

After D-Day Andrews worked on several other projects. In August he went on an experimental mission with American soldiers aboard a B-17 (Flying Fortress) to test the Eagle V camera. He also advised the Americans on some air-photography problems that they were having. He wrote a report about mapping the Normandy beaches. In a letter to his wife he said, "I have since learned, as a result, that we could have turned out better stuff, had I known what I know now."

In the fall of 1944 Andrews went to Normandy for a week to participate in echo soundings of the D-Day beaches to check their aerial photography against the actual profiles of the beaches. He found that seven were good, with differences less than two feet (60 cm); nine were fair, with less than five feet (150 cm) difference; and five were poor, with more than five feet difference. He felt that the results were about what was anticipated for this new use of aerial photography. Andrews believed the cause for the worst error was unsuccessful approximation of tilt. He also thought that the correlation between the lines run by the survey vessel and the profiles from the air photos may not have been totally accurate. He noted that the use of better equipment would have produced better photographs to make the maps. In a letter to Jean he described his time at Normandy: "Worked and lived entirely with the Royal Navy during the time, and you can imagine what an interesting time it was. Spent the nights, and the stormy days on shore in a little Normandy fishing village, and lived in the Senior Officers' Mess, RN.... I worked with what corresponds to a surveyor in the navy."

Andrews also spent a few days working on aerial photography in the Netherlands. He wrote to Jean, "We are in a Dutch town – and it is flat country!"

In November, in another letter home: "I guess most of us are beginning to feel that the war is lasting a long time, especially those of us who have been away from our families since the early 40s. I am more than ever glad that we were given that brief but glorious interlude together last year. It has helped so much."

At the beginning of 1945, with the end of war in Europe in sight, Andrews started thinking about his career after the war. He wrote Jean on New Year's day: "It would have to be something almost fantastically compelling to make me abandon my work in BC, especially without going back to see what sort of a thing I can make out of my obvious work there.... As far as I know now, I'm definitely coming back, and with a hell of a lot of new and useful ideas in air survey too." Over the winter he continued to work on developing his stereoplotter, assisted by the Williamson Manufacturing Company who had produced the Eagle V cameras.

After D-Day, Canadian military leaders became even more aware of the importance of aerial photography. In May Andrews was promoted to Lieutenant Colonel and sent on a solo mission to examine military surveying procedures in several war theatres around the world and to produce a report. Lyle Trorey took command of his section. In the introduction to his report Andrews mentioned that the Canadian survey services had only operated on the Western Europe front. He described the purpose of his mission:

> In addition to observing the survey problems and how they were tackled in the various war theatres, opportunities were to be taken, where feasible, to examine the civilian survey and mapping set-up in each of the various Dominions, India, and certain colonies, and to learn what plans, if any, were being formulated for post-war rehabilitation of the war-developed mapping potential in personnel, equipment, methods and organization. In general, air photographic survey and intelligence was to be

given emphasis, due to its war proven importance, and because the writer to some extent had specialized in it.

Over the next four months he went on 63 flights totalling over 200 hours. He made notes of methods, equipment, projects accomplished, training and efficiency. He met with many different officials, learning new ideas and sometimes making suggestions. Before departing he wrote to Jean:

> Figure about four to five months, and it's funny, but I feel in my bones
> that I'm starting on my return home to you and Mary. Granted, it's a little
> roundabout, but there is the luscious feeling that I'm on my way. I've been
> stuck here a long, long time. Bill and I landed in Liverpool five years ago
> this morning. I also am sure that the trip will speed rather than delay our
> reunion.

Andrews left England on May 12. His first stop was in Italy where he spent about ten days. He wrote to Jean: "It is a wonderful feeling of relief to be started on this active and fascinating trip – and to be in a land of sunshine. I feel marvellously fit, and the change is going to do me a world of good."

His report included some impressions of the effects of war in Italy. "The air view from these flights offered a good appreciation of the arduous terrain where the heavy fighting had taken place.... It seemed that every individual acre, from sea level to the highest ridges bore conspicuous witness to the intense fire." Then he added, "The writer's visit at Eighth Army HQ [British] coincided with the arrival there of the first keg of beer from a newly operating brewery in the vicinity."

A week in Egypt followed, and Andrews expressed to Jean his thoughts on that country:

> The sun beats down unmercifully – and in the mid afternoon the wind
> from the desert is like a furnace blast – hot and desiccating – gives one
> visions of dying of thirst and sunstroke in the desert. A big motive in
> Egyptian life is escape from the sun – any bit of shade – and escape from
> poverty and filth.... One looks down on the floodplain of the Nile – from
> the air – a green and yellow chequered carpet laid across the red-brown
> infinity of the desert. Each fragment of the pattern represents a different
> season – spring and harvest, side by side. Weather is not a topic of conver-
> sation. There is no weather. The freshness of the early morning – the quick
> gush of sun and heat rising to a crescendo with hot desert winds and dust
> in late afternoon – a gradual relenting – a fleeting but gorgeous sunset –
> twilight – and then the cool peace of night. Every day is the same.... You
> must forgive these philosophic reflections. The impact is a bit staggering!

While in Africa he also visited Kenya, Sudan and South Africa. He then headed north to Palestine, where he met Major E. William Nesham, who had been a member of the Geodetic Survey of Canada and worked in BC

Using a stereoscope
in the field in New
Guinea.

in 1909 on the Alaska–Canada Boundary survey. He spent some time in Iraq and Iran before heading on to India, where he met the country's surveyor general, Sir Oliver Wheeler, the son of A.O. Wheeler, an important BC surveyor. Then, in Ceylon (now Sri Lanka), he met Major W.W. Williams, the Cambridge physics professor who developed the mathematical formula for calculating the depth of water from wave velocities that Andrews had used in mapping the Normandy beaches. From there, Andrews flew on an American B-24 Liberator to Perth, Australia – at that time the 19-hour flight was the longest airplane trip in the world. During his time in India and Australia, he learned how the monsoon season limited opportunities for aerial photography.

Andrews spent a couple weeks in Australia, where he heard that Canadian Military Headquarters in London had approved his proposal to visit Hawaii, and then take 30 days leave in Victoria, before going on to Ottawa. "I could have roared out my joy to all and sundry – as it was I could only stand drinks to a nice old major who is my host here."

Andrews spent three days in New Guinea at a jungle headquarters where he observed how heat and humidity in the tropics affected survey equipment and maps. Then he arrived on August 19 at Biak, an island off the northeast coast of Dutch New Guinea.

This is sort of an air staging point, now have to wait a chance for an
onward lift toward Manila. The camp is on a coral island – and the blue
tropical sea is just a stone's throw in front of my tent. Here and there is
a palm tree with its feather mop top held up in the sky by the smooth
trunk. There were many more not so long ago, but the tops have been
shot away by heavy naval fire, leaving only a jagged end prong. There was
very heavy action here, and many a lad from the USA came ashore, never
to leave. It makes me feel very humble to come to such a place after the
others have made it safe.

Andrews was in both Guam and Hawaii for VJ Day, the end of World War II in the Pacific. On September 9, after almost four months of travelling, he arrived in San Francisco and proceeded to Victoria for a few months of well-deserved leave with his family while he worked on the report of his mission. He spent January and February 1946 in Ottawa, with a week in Washington, DC, where he visited the US Army Map Service and other organizations. In March he was demobilized in Vancouver.

Andrews' war experience provided him with an opportunity to expand his knowledge of aerial photogrammetry and contribute to winning the war. "My six years in uniform, like the five years between high school and university, were not lost time. After the initial frustration, my war service was a veritable post grad program. Providence caused many wonderful people, both civilian and military, to inspire, enlighten and help." Although he enjoyed living in another country for five years, he missed British Columbia and his family. It was particularly difficult being away so long from Jean, who he had married less than 18 months before leaving, and from his infant daughter, Mary. After living with only her mother for six years, it took Mary some time to adjust to having a father she did not know become a part of her life.

Surveyor

1946

Like millions of soldiers around the world returning home after the war, Gerry Andrews wondered about his immediate future. He had served for almost six years and been away from Victoria all of that time except for a few weeks in 1943, and he had lost close contact with the Forest Branch. Still, he was eager to apply the knowledge and experience that he had gained in aerial photography during the war. While he was on leave in Victoria in the fall of 1945, Andrews visited the Forest Branch, now called the BC Forest Service. He knew that the Forest Service, as a government department, was obligated to reinstate him at his previous position of assistant forester, but to his disappointment he found that he would have to go through the traditional system of promotion by serving in district offices, he would not receive an increase in pay for his war experience and he would not resume the aerial photography work that he was doing before the war. Mulholland's prediction of what would happen to Andrews after his time in the military had come true.

While he completed his report on his international mission, he spent some time exploring his options for future employment. He considered remaining in the army, going into private forestry consulting, teaching university, and conducting aerial photography for organizations outside BC. But he really wanted to do aerial photography for the provincial government. In a letter to a forestry colleague he wrote:

> The trouble is that the FB [Forest Branch] under present leadership
> doesn't recognize the fact that air survey and air-photo intelligence is
> much bigger than just a minor adjunct to the old time forest survey, and
> that it calls for a brand new appointment of an air survey engineer, equal
> in grade to a district forester, with a staff in keeping with such a grade. I

guess as far as the FB is concerned, I've had it. However, I haven't given up the conviction that the BC government must have an interdepartmental air survey and air photo intelligence service to handle not only the FB's requirements, but those of all the other departments.

Fortunately, George Melrose, who Andrews did the lookout photography project for in 1936–37, was now the deputy minister for the Department of Lands, and he supported Andrews' work in aerial photography. If the Forest Service was not going to have him continue his program, Andrews would get the Survey Branch to establish an Air Surveys department. In addition, the minister of Lands was E.T. Kenney, the MLA from the Terrace area who Andrews had met and flown around his riding when he was doing aerial photography in that region in 1938. F.C. Green, BC's surveyor general during World War II, also supported the idea of Andrews developing an air-survey program in his department. Green retired in 1946. The new surveyor general, Norm Stewart, had headed the survey section of the PGE Resources Survey in 1929 and was also a strong advocate for aerial photography.

In late January 1946, Melrose wrote to Andrews:

> The situation with regard to the position of Air Surveys Engineer is a little more clarified in that the Comptroller of Expenditure has left it in the estimates without any question. This means to all intents and purposes that it will go through.
>
> However, being a new job there is always the bare possibility that it won't. I cannot quite envisage this as the Minister has approved it, and it is merely a new job for you and not bringing in extra personnel.
>
> The next move, of course, will be up to the Premier when he goes over the estimates and finally the Legislature. I doubt very much that there will be any difficulty but you know the odd hazard as well as I do. I shall keep you in touch with any further moves.
>
> Please let me know if you hear anything that might influence our Air Surveys program, and also when you might be getting out of the army.

Because Andrews' position had been held for him and the work was similar to what he had been doing before the war, the government simply needed to transfer it from one department to another. Although he was reluctant to leave the Forest Service, Andrews knew that transferring to the Survey Branch was the only way he could continue developing an aerial photography program with the BC government. The Forest Service was one of the main departments that used aerial photographs and Andrews wrote in later years: "By accelerating the basic mapping of BC and the wide use of air photo intelligence, I was able to contribute far more to forestry and other resource developments than would have been possible had I followed an orthodox career in the BC Forest Service."

Poster from the Department of Lands and Forests.
(From *BC Government Sessional Papers*, 1950, volume 2.)

Andrews had some fortunate circumstances in developing the post-war Air Survey program:

> The batch of superb Eagle V air cameras which could and should have done the wave velocity photography so well, became "war surplus" finding their way to storage in the cavernous cellars under the Supreme Court building in Ottawa. As the RCAF showed no interest, they were consigned to "War Assets" for disposal. By some agile manoeuvring in Victoria, they were quickly purchased as a job lot by the BC government for post-war mapping and forest inventory. Between 1946 and 1960, when becoming obsolescent, they had photographed more than 400,000 sq. miles [1,036,000 km^2] on some 250,000 air negatives.

> Three of the wartime crew, Duff Wight, Howard Davis and Bill Hall joined me in the BC Surveys and Mapping Branch....

> With such improved field facilities, war surplus government-owned and modified photo-aircraft (Anson Vs) and superb war-surplus Eagle V air survey cameras with accessories, and keen young war-trained pilots, mechanics, navigators, photographers, compilers and photo-interpreters, our post-war efficient operation soon got into high orbit. Instead of the 1936–40 average of some 4,000 square miles [10,000 km^2] photographed per season, the five year seasonal average, 1946–50 was some 24,000 square miles [62,000 km^2] of superb photography at unit costs far below the nearest competitive prices.

During World War II the government had done almost no aerial photography except for the Alaska Highway so "there was a grievous backlog of mapping BC's rugged terrain and stock-taking of her natural resources, which had been at a standstill during six long war years." Aerial photography had improved considerably during the war years. There were better optics and cameras. New airplane models enabled the camera to take better quality photographs. In BC improved airfields had been constructed in many locations around the province to accommodate the larger, more powerful airplanes that replaced the float planes for most of the aerial photography work.

In his 1946 government report Andrews described the structure of the Air Surveys program:

> With the appointment of the writer as Air Survey Engineer, on the staff of the surveyor general, March 1st 1946, all provincial government air-survey activities were consolidated under his direction. Governing policy and yearly operational programs are formulated by the Interdepartmental Committee on Air Photography, under the chairmanship of the Deputy Minister of Lands. All provincial government departments and services interested in air photography have representation on this committee, authorized by the Premier on December 15, 1945. Coordination is

maintained with a similar federal committee at Ottawa, so that operations by the Province and those by the RCAF in the Province will dovetail to avoid duplication of effort. The Air Survey Engineer is technical adviser to the interdepartmental committee, and is charged with carrying out its programs within the limits of personnel, equipment, moneys and other facilities provided.

In his aerial photography work before the war Andrews had observed how other departments became interested in his program and saw how aerial photographs could assist their activities. The post-war Air Survey program would coordinate all of the province's aerial photography work and provide an opportunity for interested departments to have representation on the committee which Melrose chaired.

In his government report Melrose outlined the importance of the new department Andrews headed:

> For the first time the surveyor general had at his disposal the services of highly trained and professionally competent men in the Air Surveys Division.... Basically designed for full topographic coverage, the results of its work are imperative for modern land, forest, geological, hydrological and utility development. Every effort is being made to keep the standards as high as any in the world through the use of the best instruments by the best-trained men.

The long-term goal of the Air Surveys program was to provide aerial photographs for the entire province, but there were many immediate needs. Since Andrews was not able to begin until March after his discharge from the army, there was little time to prepare for the 1946 field season. He had to hire and train personnel and become familiar with the new equipment. "The purchase of this sorely needed equipment was completed in June, and it arrived in Victoria during July. Some minor modifications were rushed through to adapt the new cameras to our peace-time requirements in British Columbia, and a special camera mount of novel design was improvised." Andrews had not yet been able to arrange for the airplanes. "However, at the last moment a newly formed Vancouver company was able to offer an Anson V aircraft on charter, which could be ready almost immediately for operations." The aerial photography program for 1946 was established in June, with the RCAF handling the majority of the work for the first summer.

In his report, Surveyor General Stewart wrote:

> The Forest Service was anxious that a large program of air photography be carried out for a revival of the forest surveys program, and had funds available to finance it. The areas chosen were of general interest to other departments, so it was decided to go ahead on that basis....
>
> [The] 1946 agenda for air photography in British Columbia, adopted

jointly by the federal and provincial governments as on June 10th, embraced a total area of over 30,000 square miles [78,000 km²], of which the
RCAF, with several aircraft, were to attempt about 20,000 square miles
and the Air Surveys Engineer would attempt 10,000 square miles with
one aircraft.

(The total area of British Columbia is 364,764 square miles, 944,735 square
kilometres.)

The initial season for the BC government's Air Surveys program was
highly successful, and they were able to photograph almost double the
amount of territory.

Operations on first priority projects on the coast began on June 19th and
were completed on July 22nd, using Patricia Bay and Comox air bases,
during which period over two weeks' bad weather was utilized to install
long range tanks in the aircraft. Work from Prince George base began
August 3rd and finished September 10th. A number of miscellaneous
second-priority jobs were done from Vancouver airport September 18th
and 19th, winding up the season's operations for the year.

Over two-thirds of the area covered in the 1946 field season was work
done from Prince George. Much of the aerial photography that Andrews'
department did during the first years was in northern BC because of the
numerous resource development projects that occurred in that part of the
province after the war. Andrews reported:

The area covered is roughly three times that done with one aircraft in the
average pre-war season, and, in spite of higher general operating costs, the
unit cost (per square mile) is the lowest on record. It is felt that this is a
very gratifying result, in view of the many difficulties and uncertainties
which had to be met in this first year of post-war Provincial air survey
operation.

He believed that there were several factors that contributed to the success of the initial season. "Large authorized area of operations from one base
permitted full exploitation of weather opportunities." The higher performance aircraft permitted better coverage of an area and a longer range of
operation. This enabled the air photography crew to cover a larger area from
a single operations base. Improved weather reporting and communication
allowed for more efficient planning and use of resources. The "advanced
optical features, precision design, and construction of the new Eagle V cameras" produced quality photographs. There was also

the conspicuous success of our new home-made camera mount for completely insulating the camera from vibration, and for minimizing the tilt
effect from haphazard lurches of the aircraft in flight. This permitted the
use of a slower shutter-speed and smaller aperture in the camera, which
enhanced definition in the photographs without jeopardizing exposure.

[Most important was the] high standard of effort and discipline in the operational personnel, achieved by careful selection, training, and attention to factors contributing to good morale.

Once the work was done in the field, the aerial photographs had to be processed and catalogued. Initially the Air Surveys program used the photographer from the King's Printer, but he was unable to keep up with the volume of pictures. "The present volume and the specialized nature of our air survey processing now makes the provision of our own dark-room facilities imperative. An initial step in this direction has been taken recently in the setting up, in temporary quarters, of a new projection printing unit and the accession to our staff of an expert on processing." At the beginning Andrews supervised the provincial air photo library.

A new series of standard indexes of air photography for the Province has been initiated, at a scale of 4 miles per inch, and following map sheet boundaries of the national topographic series.... The air photo library now has approximately 150,000 air photographs in its collection, and a better system of filing them has been provided by the acquisition of sufficient steel filing cabinets from War Assets to accommodate 200,000 photographs.

From the beginning of the program Andrews wanted to ensure that the public also had access to the aerial photographs so people could see the value of the work of his department. "Grade N Photostat negatives of the up-to-date manuscript copies of the sheets will be kept on file, from which inexpensive blueprints may be taken to meet demands from other government departments and from the public."

He summarized the first season of the program and then looked ahead to the future:

The lack of adequate base facilities on the ground at the head office is now the weakest aspect of our air survey picture. The accelerated appreciation of air photography in connection with the post-war development of our Province, evidenced by the Interdepartmental Committee on Air Photography and the success of our flying operations recently completed, justifies the final step of consolidating the work into an officially recognized Air Survey Division of the Surveys Branch, with suitable accommodation, installation of special equipment, and adequate personnel.

The Easy Eye ready for take-off at Smithers.

1947

The Air Surveys program expanded to a full division of the Survey Branch in 1947, although poor weather limited the number of flying hours. This season Andrews had two airplanes available for aerial photography with the funds for one of them coming from the Forest Service. One airplane continued the vertical aerial photography, while the second (CF-EZI, affectionately known as the Easy Eye) started a tricamera photography program. Andrews described the special mount for the three cameras that was installed in the airplane:

> The tricamera mount successfully embodied the desirable feature of maintaining the three cameras rigidly in constant angular relationship, provision for levelling and rotation of the camera assembly as a unit, for normalizing drift (or crab) in flight. Its design employed the three-point floating suspension device originated by us last year in connection with vertical camera mounts, and which proved eminently successful for dampening vibration and for smoothing out sharp tilt effects arising from haphazard lurches of the aircraft. The arrangement of the cameras in the mount followed the standard practice of oblique camera axes depressed 30 degrees below horizontal, to form an angle of 60 degrees with the vertically directed axis of the central camera.

He noted that "this installation was possible only with cameras of small size such as an Eagle V".

In the tricamera method, the middle camera took a vertical photograph while the two side cameras took oblique pictures. This enabled both a direct view of the land below and the adjacent terrain. Andrews believed that "it is possible to expect that when good tricamera and basic vertical photog-

A tricamera installation. The three Eagle V cameras swing as a unit. (Aero Surveys photograph, from *BC Government Sessional Papers*, 1948, vol. 2.)

raphy are combined, the accuracy will be sufficient to make detail maps at as large a scale as 40 chains [half a mile] per inch, suitable for preliminary extensive economic purposes." By being able to view a larger area than that covered just by vertical photographs, it would be possible to begin looking at the potential for a variety of project developments over a wider area of the province, particularly in the more remote northern half of BC.

The tricamera method was used to start a new project in 1947, the Western Highway. The Alaska Highway was the only major road in northern BC. Its route went through the northeastern part of the province, then across northern BC near the Yukon border. There was no road north of present-day Highway 16 in the western part of the province. Minister of Lands E.T. Kenney, the MLA for Terrace, believed that a road through northwestern BC would give access to a large forested and mineralized area and open the region to economic development. It would also provide access to the ocean at Stewart. There were three potential routes: the Rocky Mountain Trench, which had been mapped during the original Alaska Highway survey; the Middle Route, which went from Vanderhoof to Fort St James, then northwest up the waterways beyond Stuart Lake; and the Western Route, which was closer to the Coast Mountains. The Western Route had three possible starting places: Terrace, Kitwanga or Kispiox (initially the preferred choice). The tricamera method was used to photograph both the Middle and Western routes. The vertical photograph followed the route, while the oblique photographs showed the nearby terrain, enabling the surveyors and engineers to assess potential difficulties and examine a variety of alternatives.

In 1947 aerial photography covered the potential highway north to Telegraph Creek, and the following year it continued to Atlin. Surveyor General Norm Stewart wrote about the project in an article for *BC Professional Engineer* in 1954:

> From the accumulated data, the air reconnaissance, and the air photos,
> a photographic report was made. Stereoscopic pairs of oblique photos
> were selected and on them the route was indicated with arrows pointing

The Mount Edziza area of the proposed Western Route, showing the prominent features and compass direction. The wing of the plane is visible in the corner of the picture. Andrews took a series of reference pictures while doing the aerial photography.

to many features of interest. This photographic report provided a quick means of viewing the whole route. It also proved very useful in preparing instructions for the mapping program which got under way at Hazelton in 1949.

Before World War II Andrews had struggled to get an aerial photography program established with the BC government. In the years after the war, the technology's popularity grew rapidly. People wanted to use the information that could be gained by looking at the earth from the air. Andrews wrote:

The larger agencies, both government and private, are generally well versed in the use of air photos: they know exactly what they want, see quickly what is available, and order their requirements with a minimum

Powell River vertical photograph, July 17, 1947. When viewed with a stereoscope, aerial photographs of a community would show a three-dimensional view of the houses and the industrial use of the harbour.

of trouble. There is, however, a large and growing class of "lay" users who have had little or no previous experience with air photographs....

At the close of this year we have a formidable backlog of printing orders, still to be done, for the air photo library, Government departments, and for the public, of some 28,000 prints. There is no sign of this business letting up, rather it is steadily increasing.

[For viewing these photos] another batch of fifty mirror stereoscopes was produced by the Air Surveys Division, using local manufacturing facilities in Victoria. This 1947 model is similar in basic design to former models, but is much more compact, about half the bulk while retaining the same field of view. It has plastic panels in the body construction to permit better illumination of the photographs, which also improve its general appearance. A simplified device is provided for truing up the mirrors into parallelism, with a few seconds of arc. A novel arrangement of the mirrors reduced the viewing distance eye to photo, for closer observation and slightly reduces the size (cost) of the mirrors without lessening the field of view. These instruments cost $40 each in a carrying case. They have all been disposed of to various Government departments, and a small number, on special request, to industries which experienced difficulty getting imported equipment. Requests for additional units are already accumulating, which may justify the production of another batch early in the new year.

To continue publicizing aerial photography, personnel in Andrews' department gave four public presentations and Andrews wrote a magazine article. His aerial photography crews also started photographing towns and cities that were in their area of work. In a presage to Google Earth, Andrews commented on information about a family that could be obtained from aerial photographs: "We note in low-altitude photographs of residential areas, taken other than on a Monday, that each home where a 'blessed event' had recently occurred is conspicuously earmarked by the daily wash of white napkins strung out behind the house."

The Andrews household experienced their own "blessed event" in March 1947, with the birth of a second daughter, Kris. It was a good year. Mary, now seven years old, was especially excited to have a baby sister.

The Air Surveys Division grew that year, too. By the end of 1947, it had a staff of twelve, including three air-photo specialists. This rapidly expanding department needed additional space both to accommodate the people and the equipment needed to process and interpret the large volume of photographs being produced. In his report Melrose wrote: "Good progress was made in 1947 in the surveys of the province, especially in aerial surveys, which progressed faster than others."

1948

Andrews' organization had a sizeable increase in responsibility and personnel in 1948 when the Forest (Air Photo) Base Maps Section was transferred in from the Forest Service. As part of the transfer the Air Surveys Division had to increase their rate of aerial photographs for forest survey base maps from 3000 to 9000 square miles (7,770 to 23,310 km^2) per year by 1950. Andrews now supervised all of the BC government's aerial photography. The Forest Service transferred 18 people to Andrew's department, including Bill Hall, his long-time friend who took on a new appointment as assistant chief engineer, and Axel Kinnear, who had worked with Andrews in aerial photography before the war. Kinnear became a photographic analyst, and Andrews was glad to add his valuable experience and expertise to the department.

With 39 people in the Air Surveys Division, Andrews now had to expend a "large amount of time, energy and ingenuity ... to build up increased staff organization". This division of the Survey Branch was unique for the variety of positions involved in managing the technology of aerial photography: airplane pilots, photogrammetrists, air photo analysts, film processors, a man who made and repaired instruments, a photo librarian and a variety of air survey technicians. Andrews had many talented and innovative people working in his department who were able to keep up with new advancements in this rapidly evolving field of work. He also encouraged the professional development of his staff. Two pilots and one photogrammetrist from this group went on to become BC Land Surveyors.

The air photo library moved to larger quarters and more floor space was added for processing the film, but Andrews' department struggled to keep up with demand for aerial photographs. He noted that the "interest in air photography of this province is increasingly lively and versatile".

Field operations in 1948 were similar to the previous year with two airplanes being used, one for vertical photographs, and one for tricamera operation. The tricamera unit proved its value early in the season during the historic Fraser River flood:

> It suddenly became evident during the weekend of May 29th and
> 30th that the flood situation in the Lower Fraser River valley had reached
> emergency proportions. The value of an air photographic record was
> realized, and a tricamera flight from the Strait of Georgia to Hope was
> therefore made the next day, Monday, May 31st. This flight was at 17,000
> feet [5200 m] above sea level to try to include most of the river in the
> central vertical camera. A full set of these photos, 176 in all, were dis-
> patched to headquarters of the British Columbia Flood Control Authority,
> Vancouver, on June 9th. Three succeeding tricamera flights of the same
> area were from 10,000 feet [3000 m] on June 5th and 27th and July 14th.
> Photos from the flight on May 31st showed the waters pouring through
> the Matsqui dyke, which had just broken earlier that day. A comparative
> view exposed June 5th shows complete inundation behind the same dyke.

George Melrose also dispatched Andrews in early June to photograph some places in the interior that were flooding. Andrews wrote:

> The value of the tricamera method of photography for emergency
> cover was well demonstrated in the above instances. To have covered
> the areas exclusively with vertical photography would have involved

Fraser River flood, Matsqui break, May 31, 1948.

approximately three times as much flying, and in consequence would have increased the adverse odds for weather opportunity by the same multiple. In other words, the tricamera method enabled the emergency to be "contained" by air photo record in the very brief opportunity which our climate somewhat grudgingly offers. It might have been well-nigh impossible to have finished the job at all, while the floods were at their peak, with vertical runs, due to weather interference.

The oblique photographs enabled the entire valley to be viewed, so that people could see the total extent of the flooding. "The flood pattern, by nature a level feature, is comparatively easy to map from obliques." In a later article Andrews wrote that, "A primary virtue of our government photo flying facilities was flexibility and capability to cope promptly with unforeseen emergencies without formality or red tape, so well demonstrated in the Fraser River flood crisis of 1948." He noted that the series of air photos showed both the maximum flooding and recession of water.

The tricamera operation included the second half of the aerial photography for the proposed Western Highway. Andrews went to Atlin for the first time that summer, and during his work he met Harper Reed, a longtime resident of the area who became a close friend. There were two low altitude tricamera flights for the Water Rights Branch. In the Okanagan valley the first attempt was made to photograph the mountain snowpack to try to predict flood potential in the spring. Unfortunately the weather did not cooperate on the days they were scheduled to fly. "A flight of unusual

Fraser River flood: a southerly view of the Matsqui Prairie showing full extent of
inundation on June 5, 1948 .

interest was made along the British Columbia-Alaska boundary from Tarr
Inlet (Glacier Bay) to the extreme northwestern corner of the province....
Photographs of this flight reveal the geography of one of the least-known
parts of British Columbia and are classics in the delineation of glaciers and
ice drainages."

The RCAF covered 163,000 square kilometres in their aerial photography, much of it in the northern part of the province near the Yukon border.
Andrews observed that with the large amount of work done in the past
three years aerial photography had advanced appreciably ahead of the mapping that would be made from these pictures. One of the spare Eagle V
cameras made a trip to the Arctic islands to record topography there in connection with an expedition sponsored by the federal Department of Mines
and Resources. In 1948 colour film was used for the first time, but Andrews

The Great Pacific Glacier in northwestern BC near the Alaska boundary.

noted that more restricted light and weather conditions were needed to produce successful photographs.

Aerial photographs were made of 16 more communities, including a special one of Victoria:

> At noon, June 8th, a special mosaic of the Victoria metropolitan area was made from 5,000 feet [1500 m] above sea level at an extreme low tide for the year. On every alternate strip of this project, tricamera obliques were synchronized with the vertical camera, and on the other alternate strips a second vertical camera containing infrared film was synchronized with the routine vertical camera which contained standard panchromatic film. This project provided a new detailed mosaic of our capital city, with a maximum of offshore hydrographic detail of interest especially to navigators of small craft. It provided material for comparative studies between infrared and standard panchromatic emulsions. It also yielded an excellent set of scenic oblique views of the area.

The variety of projects undertaken in 1948 and the rapid growth of the Air Surveys Division testified to the growing value and importance of aerial photography in British Columbia.

One of the first surveying operations in BC to use helicopters took place in the mountains near Chilliwack in 1948.

1949

In 1949 the Air Surveys Division expanded its equipment and production services. In his annual report, Andrews stated that the department now had a "staff of almost fifty technical personnel organized into six functional subdivisions, 9,000 square feet [about 800 m²] of specialized floor space in Victoria, considerable inventory of photogrammetric equipment, three motor vehicles, two photographic aircraft, a share of the government-leased hangar at Patricia Bay airport, and an imposing aggregate of work accomplished." One of the biggest acquisitions of the year was a pair of airplanes. The Ansons had proven their value in conducting aerial photography and the government decided it was cheaper and provided more versatility to purchase their own aircraft. One of them was CF-EZI, which continued to do the tricamera photography.

Andrews reported on another important development:

> We are very proud of the large room especially built for slotted-templet
> lay-downs, a floor space of nearly 2,000 square feet [almost 200 m²] with
> trussed roof free from pillars, for laying down large areas of country in one
> mapping project. This is the first instance we know of (in Canada) where
> special construction of a government building for this purpose has actually
> been won and done.

The purchase of a six-projector multiplex unit allowed the department to make air mapping even more precise. Andrews later wrote:

> When the six multiplex units were finally installed and operational in
> special premises our Minister, Mr Kenney, and Deputy, George Melrose,

thought it would be a good idea to have the Premier, the Hon. Byron Ingemar ("Boss") Johnson come for a demonstration.... The demonstration went off well and Mr Johnson was quite impressed. Just before leaving the Premier looked at me with a twinkle in his eye and said, "Where did you get the money to buy and install all this?"

Andrews concluded his report by stating:

With the final step of acquiring and operating our own aircraft, with experienced air crew and darkroom staff, with speedy follow-up in the office checking and indexing, we have the air photo situation well in hand. We may now expect a routine sustained production of excellent air survey photographs, in large or small quantities, wherever and whenever required, to a wide range of specifications, at extremely low unit cost....

The last two years have seen the total proportion of the province area covered by vertical air photography suddenly increased from roughly 30 to 90 per cent. This aggregate is due, in minor part, to our own provincial program, geared more or less, to a steady 10 per cent of the province area per year. The major cause of the jump is an inordinate stepping up of federal photo flying in this province. In 1948 alone the Royal Canadian Air Force covered 63,000 square miles [163,000 km²], an area almost equal to their previous cumulative aggregate for the two decades since the inception of operations about 1926. This year the Air Force doubled its 1948 figure by doing 126,000 square miles in British Columbia.

The RCAF had as many as nine modern airplanes doing aerial photography in BC during the summer of 1949. In his report Deputy Minister George Melrose summarized the importance of aerial photography: "Air photos are basic to knowledge of surface features, which in turn permits resources inventories, provides data for the preparation of maps, and supplies indispensable information for planning the development of the resources of the province." The Forest Service remained the major government user of aerial photography throughout BC. Melrose reported:

Good progress has now been made by the Forest Service in setting up air photo libraries at each district office ... and smaller libraries in each of the ranger districts.... All the district offices are now equipped with stereoscopes and the rangers are being supplied as opportune. Senior members of this Division have conducted short courses in the fundamentals of photogrammetry and interpretation at district ranger meetings and at the Green Timbers Ranger School.

The Water Rights Branch used aerial photography for several projects. Since the end of World War II, it had been examining many potential sites for hydroelectric and flood-control dams to go with the development of the province's natural resources and growing population. Initially the branch looked at several possible sites along the Fraser River, with inter-

Slotted templet laydown. (From *BC Government Sessional Papers*, 1949, vol. 2.)

est increasing after the 1948 flood. In 1949 Ernie McMinn surveyed the area around Kitimat for the potential Alcan aluminum project. His intent was to establish ground-control stations for aerial photography and use the slotted-templet method to create a map of the area. Aerial photography was also done around Chilko Lake, a region that the Water Rights Branch had examined before World War II. The Air Surveys Division successfully conducted two aerial photography flights to sample snow for the spring runoff.

Andrews' department also provided aerial photographs for the Pacific Great Eastern Railway to assist in the completion of the last section of the rail line to Prince George, and to begin planning the extension of the railway north to the Peace River district. Approximately 300 kilometres of the most northerly section of the Alaska Highway was photographed, both for road improvements, and for right of ways along the highway, as use of the road increased. In addition, "two experimental flights were made to photograph salmon in their spawning grounds – one at the north end of Chilko Lake and the other on the Stellako River. Both flights were at an altitude to give a net height of 1000 feet [300 m] above the rivers. These attempts were hampered by long shadows and high winds over the waters." Despite the lack of success this time Andrews believed that "a large field remains for experiments such as this."

The large amount of aerial photography in 1948 and 1949 created a backlog in mapping. In his report, Andrews commented on Air Surveys Division's attempts to keep up with the huge volume of new photographs:

> We are not yet by any means satisfied with our mapping output, but we are at last arriving at a point where something can be done about it. The air photography side, which has consumed a large proportion of our available energy, should now be largely self-propelled, so that extra power may be put into the mapping work.

1950

In 1950 the Ministry of Lands reorganized the Survey Branch and renamed it the Surveys and Mapping Branch. Norm Stewart remained the director of the organization but relinquished his duties as surveyor general to Frank Morris, who headed the Legal Surveys Division. Andrews became the assistant director of Surveys and Mapping while retaining his position as head of the Air Surveys Division. He now became involved in the operations of the entire organization.

This year the RCAF photographed about 52,000 square kilometres, primarily in two blocks: the northeast corner of the province, and the northwest in the Alaska Panhandle area. Combined with the aerial photography done by the provincial government, the initial aerial photography of BC was completed in 1950 after five busy years in the province. On November 19 an editorial in the Victoria *Daily Colonist* reported:

> RCAF planes, acting in concert with the Lands Department, have now completed the aerial mapping of British Columbia once over, in some 300,000 photographs taken at a uniform height of 16,000 feet [4900 m]. The basic photographs must be supplemented by others and considerable detail added before cartographers can produce the maps from which every known facet of the province can be studied in years to come. It is an achievement, however, to fit British Columbia into a set of photographic plates....
>
> [This is the] completion of a task that will probably have a greater bearing on the future of the province than all the rush and excitement of front line news for five years back.

The Forest Service remained the main user of aerial photographs, but the Deputy Minister of Lands observed:

> the work of the Water Rights Branch for the past year approached or reached new records in several respects.... Further progress was made in the investigations of the Fraser and Columbia watersheds in co-operation with the governments of Canada and the United States, and a large number of other investigations were carried out during the year. Among the most notable water licenses issued was on the Nechako and Nanika rivers for the development of power for the proposed multi-million dollar aluminum industry.

The Air Surveys division provided aerial photography for these projects.

The provincial government's surveying department, particularly the Air Surveys division, was also involved in several other natural resources projects. In addition, "to provide for the tourist and the vacationist, the province has disposed of a large number of sites along the Alaska Highway for service stations and tourist accommodation. On various lakes or at other suitable

A legal survey monument post at Bowser Lake, next to the historic gravesite of Simon Gunanoot. A member of the Gitxsan First Nation, Gunanoot was accused of murdering two "white men" near Hazelton in 1906. He eluded authorities by living in the wilderness with his family for 13 years before surrendering on his own terms, then stood trial and was acquitted in 1920. Gunanoot died in 1933.

locations, lands have been acquired for the building of hunting and fishing lodges and dude ranches."

In the early part of the 20th century the Alberta and BC governments had surveyed and agreed on the boundary between the two provinces for the Rocky Mountain section and a portion of the 120th meridian. The northern boundary along the 120th meridian from 57°20' to the 60th parallel was not completed because the area was remote, difficult to access, and not considered economically important. After World War II the growth of the oil-and-gas industry in northeastern BC and northwestern Alberta made it necessary to delineate the 120th meridian up to the 60th parallel. Both provinces wanted to be sure that they were receiving the revenue from all the resource sites in their region. In 1950 the Alberta-British Columbia Commission was re-established and the BC government completed a tri-camera photographic survey of the BC-Alberta boundary along the entire 120th meridian.

Andrews participated in a history-making project in 1950 – the first completely airborne topographic survey in British Columbia. This survey used conventional airplanes and a helicopter to photograph a difficult area for ground travel in the vicinity of Meziadin and Bowser lakes. It was part of the ground survey for the Western Highway, and a continuation of the work that Andrews had started in 1947 with the aerial photography of the

Mount Pattulo, during the triangulation survey of northwestern BC's rugged mountains using only aerial transportation. The BC government produced a film called *Flying Surveyors* that chronicled this pioneer surveying project in the province.

potential route. He assisted in the organization of the project and inspected the surveys in the field. This air survey mapped the area without the need to blaze trails, employ axe-men and mountain climbers, pack equipment on horses, or establish numerous base camps.

Now that Andrews was in one of the top positions in the Surveys and Mapping Branch, he believed that it was important to become a registered BC Land Surveyor. Before World War II he had articled to Frank Swannell but had been unable to work in the field for him. In June 1939 Andrews had written to J.R.C. Hewett, president of the BC Land Surveyors, requesting permission to take the final exams. He stated that he still wanted to be a surveyor and that his work in aerial photography was "closely identified with one of the broader and more modern aspects of surveying". He also noted that "preference and opportunity has gravitated toward the surveying side of forestry engineering". Andrews asked if the board of the BC Land Surveyors would consider his experience equivalent to articling. This was an unusual request, because he had not followed the traditional role of spending time in the field working on a survey crew with a BC Land Surveyor, but it was a reflection of changing times and new technology. The board decided to accept Andrews' request if he would complete the diary forms and if Frank Swannell would sign his discharge from the articles. As one of the premier surveyors in the province, Swannell's assent was good enough for the board and Andrews could write the exams. In early 1940 Andrews and Swannell agreed to the board's conditions, but by then Andrews had enlisted in the army. In July 1940 Andrews received word that he could write the final exam when he returned from military duties. Then, after the war, the Land Surveyors Corporation changed the regulations and the registrar ruled that Andrews was not eligible to take the exam. Now, in 1950, the new secretary-registrar discovered board minutes from 1940 stating that any BCLS pupil who served in the war would be eligible to write the final exams. Andrews passed the first part of the exam in April 1951 and the second part in the following spring, becoming BCLS #305.

1951

In 1951 the Surveys and Mapping Branch reorganized again. Early in the year Norm Stewart retired as director. Andrews was supposed to assume this position, but he wanted to remain head of the Air Survey division longer to complete its organization and to have time to become familiar with the director's job. Frank Morris added the director's responsibilities to his, but then, in late spring, he also retired, so Andrews became the director of Surveys and Mapping and the surveyor general.

During that summer Andrews visited all of the survey parties working in the field so that he could become more familiar with the projects and get to know the surveyors and the procedures they used in their field work. Almost every summer thereafter Andrews would leave Victoria for two or three weeks to visit as many of the crews in the field as possible. He enjoyed travelling around the province and meeting the survey crews. The outings also provided an opportunity to visit old friends and familiar places en route. He usually brought a couple of his Victoria staff on these trips, different people each year, usually a senior and a junior employee. A flask of overproof rum in a red wool sock was a standard feature of an Andrews visit. His friendliness toward all the people he met in the field, whether registered surveyors or summer staff, and his genuine interest in the work they were doing fostered camaraderie with his employees. Andrews developed a similar rapport with his office staff in Victoria.

In his annual report Andrews wrote:

> The main theme for 1951, recurrent almost monotonously throughout, is the pronounced increase in all phases of the work, resulting in part from accumulated training, surveys and mapping service which we are set up to provide. In preparation of this report we have indeed been hard put

Andrews in his Victoria office in conversation with secretary Beth Albhouse.

to find alternatives for the word "increase", having long since exhausted Roget's store of synonyms.

A year ago we thought, almost hopefully, that our curve of post-war expansion was about to flatten off, that we could direct some effort to consolidation and, perhaps, to a bit of polishing up. However, surveys and maps being so integrated with development, in fact being the very "blue-prints" for it, we find ourselves carried along involuntarily with the phenomenal tide of industrial expansion now sweeping this Province, witnessed by such names as Columbia Cellulose, Hart Highway, Pacific Great Eastern Extension, Alcan, Duncan Bay Pulp, Celgar Pulp, Peace River Oil and Trans-Mountain Pipeline.

One of the Surveys and Mapping Branch's major projects was surveying related to the construction of the Alcan aluminum plant at Kitimat. In 1951 Art Swannell completed a triangulation control survey of the Tweedsmuir Park area (the northern part was going to be affected by the flooding of the upper Nechako River behind Kenney Dam) and tied this in with the tunnel survey of the Alcan project at Tahtsa Lake. Frank Swannell had made the first detailed surveys of the upper Nechako waterways and Nanika River between 1920 and 1925, so Art was following in his father's footsteps. He surveyed the adjacent Nanika River, also part of the project, in 1952. In the spring of 1952 Andrews attended the official ceremony related to the construction of Kenney Dam (part of the Alcan project) south of Vanderhoof and Frank Swannell accompanied him.

Some of the area that
Art Swannell surveyed
for the Alcan project.
(From *BC Government
Sessional Papers*, 1952,
Volume 3.)

Triangulation Survey

Tweedsmuir Park Area

Beaver aircraft,
Tesla Lake.

Coast
Mountains.

Tesla Lake and
Rhine Crag.

In 1951 Gerry and Jean Andrews purchased a home with two hectares
of land on Blenkinsop Road at the foot of Mount Douglas in Saanich, just
north of Victoria. The previous resident had been a keen amateur horticul-
turalist who created a large rock garden with many species of flowers. Jean
and daughters Mary and Kris enjoyed working in the garden. Gerry tended
a family vegetable garden, a sizeable orchard and a small plot where he grew
his own pipe tobacco. The household included the family dog, Bingo, Kris's
bantams and (for a short time) a peacock named Michelangelo. The home
on Blenkinsop Road gave Gerry a place to relax and to entertain friends
and colleagues. Like most women of her time, Jean did not work outside
the home (except briefly during World War II), so the family, house and
garden occupied her life, especially since Gerry was often away from home
or worked late. She devoted herself to keeping a beautiful house, cooking
and entertaining guests.

Jean did not like "roughing it", so the Andrews family rarely went on
camping trips. But they often took a Sunday drive around the Victoria area

Jean Andrews with daughters Mary (centre) and Kris on Mount Douglas, near their property in Saanich.

and sometimes up island. Jean and Gerry took a few trips to eastern Canada and once travelled with Kris to Europe. Occasionally Jean visited her family in California, usually by herself, but Mary and Kris accompanied her a few times.

Gerry's daughters remember their father working hard at the office and at home. He was dedicated to the work he did and the people he worked with. But when he was home he was a generous and supportive father. Jean and Gerry were both in their thirties when they married, and in less than two years the war intervened. Like many wartime couples, the long years of separation impacted their lives. But, according to Kris and Mary, both were good parents who respected each other and enjoyed a satisfying marriage that lasted more than 60 years.

1952

In his 1952 deputy minister's report, George Melrose trumpeted the successes of the Department of Lands, including the Surveys and Mapping Branch:

> Land and water developments in British Columbia during the year
> 1952 marked the greatest surge of economic and industrial activity in the
> history of the province. Giant enterprises were being carried through, par-

Surveyor A.J. Campbell (right) inspects material to be used in the building the Hart Highway near the Parsnip River bridge.

ticularly in the fields of forestry and water power, and in their wake came a multitude and a variety of applications for land. This spectacular activity has resulted in new records being set in the operations of the branches of the British Columbia Lands Service, the Lands Branch, the Water Rights Branch, the Surveys and Mapping Branch, and the Coal, Petroleum and Natural Gas Branch.

For the Surveys and Mapping Branch, he listed surveys related to the construction of the Trans-Mountain pipeline from Edmonton to Vancouver, right-of-way surveys on the Alaska Highway, the Cariboo Highway, the John Hart Highway north of Prince George and the Hope-Princeton Highway, and topographic mapping for the Water Branch along the Fraser River and in the Columbia River basin.

Andrews spent much of his first years as surveyor general establishing the boundaries of British Columbia with its eastern and northern neighbours. In 1952 he was appointed BC's representative on both the Alberta-British Columbia and the British Columbia-Yukon and Northwest Territories boundary commissions. In his report for 1952 he described some of the work done by his branch in difficult conditions. Winter work was done in the muskeg country where it was easier to travel and survey when the ground was frozen.

Surveys for the establishment of British Columbia's boundaries have continued during this year. A party under W.N. Papove, BCLS, DLS, ALS,

Construction underway for the Kenney Dam.

working during the winter of 1951–52, projected the Alberta-British Columbia Boundary northward along the 120th meridian from Hay River, latitude 58°45', for a distance of some 40 miles [65 km], leaving approximately 48 miles [77 km] to the end of this boundary at the north-east corner of the province where the said meridian intersects the 60th parallel of north latitude.

During the present winter, 1952-53, it is expected that a party now in the field under George Palsen, DLS, ALS, will complete the survey of this boundary, and subject to favourable conditions of ground and weather, the same party will turn the corner and continue westward along the British Columbia-Northwest Territories boundary, aiming to connect with a portion surveyed by [recently retired] N.C. Stewart, BCLS, DLS, during the past summer, some 75 miles [120 km] distant, at the Petitot River. This is an ambitious program. Mr Palsen plans the use of dogs and light motorized toboggans for transport in the northward direction in the first part of the winter, and the use of tractors in the westward direction from the northeast corner, where he has arranged a rendezvous with the heavier units about mid January, when the ground should be frozen sufficiently to carry their weight....

Concentration of survey effort to establish the two boundaries flanking the northeast corner of the province has been necessitated by widespread activity in exploration for petroleum and natural gas under permit in that region. It is imperative to establish, on the ground, the line of demarcation between the three Governments concerned – Canada (for the Northwest Territories), Alberta and British Columbia – in a region endowed by ancient geological events with promising petroleum possibilities, with no regard whatever for man-made political boundaries.

On January 21, 1953, the survey of the BC-Alberta boundary was completed. Efforts were then directed toward completion of the northern boundary along the 60th parallel, with the British Columbia-Northwest Territories section finished in 1954.

N.C. (Norm) Stewart at the boundary for Alberta, BC and the Northwest Territories.

1953

The post-war economic boom continued during Andrews' years as surveyor general, especially in the 1950s. The organization that Andrews headed participated in almost all of the many large projects in British Columbia. Of its growth he wrote: "The Surveys and Mapping Branch in 1953 is thus a large complicated organization compared to what it was prior to the end of hostilities in 1945. Its present size, of some 200 personnel, and character, however, simply reflect the magnitude and diversification of present-day life in the province."

One important project was the development of the oil and gas fields in the northeastern part of the province.

The Petroleum and Natural Gas Act of 1946 allowed application for and tenure of exploration permits located by map description. Soon the whole region was completely blanketed with a rather chaotic array of these permits, none of which was surveyed. The favourable results indicated that soon the permit tenures would go to lease, the survey of which on the ground was a very necessary requirement.

Oil and gas had been discovered in 1951. After completing the eastern boundary of the province it was imperative for the BC government to survey the region so that the wells being drilled could be accurately located and mapped. The only previous detailed BC government survey of northeastern

Nig No. 1, one of the first oil wells in northeastern BC.

BC had been completed by G.B. Milligan (BCLS #41) in 1914. Milligan, who had a small crew, mapped the area and examined its economic potential. In his report he noted the almost level ground gave him very few opportunities to establish survey stations, so he used mainly observations of the sun and stars to determine his locations. The virtually flat terrain limited visibility and muskeg made travel difficult, so Milligan was unable to leave many permanent survey markers for future surveyors. The surveying of northeastern BC began in 1953, headed by Ernie McMinn (who became BCLS #325 that year) and lasted for three years.

As surveyor general, Andrews coordinated this project and visited the crews in the field each summer. In his report he noted that all but one of the field parties in 1953 worked on the northeastern project (Art Swannell completed the surveying for the proposed Western Highway). The survey required mapping control and field surveys to establish permanent survey markers. McMinn described the strategy in his report. The surveys could "be made by winter or summer work, by traverses or by triangulation, using towers. Despite the adverse findings in the history of triangulation by towers in inaccessible country, it was judged to be the only practical technique to employ."

Andrews believed that it would be too expensive and slow to extend the township and range survey of the Peace River region into this area. The solution was to construct a series of survey towers high enough above the trees to be visible to other towers and far enough apart so that no more needed to be built than necessary. Permanent survey markers placed in the ground under the towers would connect all surveys in the area to this network. This would be a triangulation survey done not on the ground, but above the trees.

A test model tower was built near Victoria simply to discover any unforeseen problems in building towers of poles found on the site. A scheme of testing the accuracy of observing on helium balloons could not be carried out, failing the arrival of the equipment. The idea of aluminum towers or strut-supported masts was abandoned as being too expensive.

For the 1953 field season there were four BC land surveyors "together with twelve instrument men, a three-man helicopter crew, radio man, computer, two cooks, and twenty other men, a total of forty-three. The helicopter contract with Okanagan Air Services was for 330 hours or three months. Eight vehicles were to be used to transport men and survey equipment to the Fort St John area." The BC government purchased a Beaver float plane, CF-FHF, that was used to bring in supplies, fuel for the helicopter, and move camp to different lakes in the area.

McMinn described the beginning of the field work:

> A day-long reconnaissance of the area in the Anson aircraft on May 14th enabled the author to get a lasting impression of the country and of the survey task as related to it.... The first attempts at finding tower sites with the helicopter were ended abruptly by becoming lost and partially convinced that tower building was hopeless. However, ten towers were commenced, averaging over 80 feet [25 m] each. At this stage in June the tower building was in danger of being abandoned in favour of an attempt to traversing [laborious ground-based surveying]. The first towers, however, proved to be stable and intervisible, and the tower building went on with enthusiasm.
>
> The tower sites were chosen by reconnaissance, using the helicopter. Intervisibility was tested by hovering below the top of the slight rise selected. The height of trees, suitability of camp site and nearness of landing place were noted. The "recce" [reconnaissance] man then selected a further site some 10 miles [16 km] distant on the first visible horizon and they flew directly to it. Several sites must always be selected at the same time. In a short time the flat featureless country became fairly well known, and each tower site (marked by dropping rolls of paper) could be accurately described.

Then he described the procedure used to construct a tower:

> The towers were built by three men. Their equipment consisted of single and double-bitted axes, swede saws, hammers, 100 pounds of 4, 6 and 8 inch spikes, wire, hundreds of feet of rope and blocks, fallers' helmets, climbing spurs, and their own good spirits.... The towers were built double, with an outer platform for the observers and an inner tripod for the instrument; the two parts could not touch even at ground level.
> The most economical shape, using standing trees if possible, was a double triangle, as this eliminated one side and interior bracing in each structure. The cotton signal was erected over the tripod head and a pipe post plumbed in on the ground. Additional poles were used to buttress the tripod, but no ladders were built because the structure could be climbed as it was on the cross bracing.

Once the towers went up, the surveying could begin. McMinn wrote about some of the difficulties they encountered.

Tower station Milligan, named after G.B. Milligan, the first man to make a detailed survey of northeastern BC.

As the frost left the ground some settling was observed, especially of the legs that had been dug in. This meant that the verticality of the structure must be tested on each day of observing. High winds caused a swaying of the tripod, but the vibration caused by breezes did not affect the instrument work. Lightning and rain storms made the tower untenable. June and July are the best observing months because the northeast and northwest light of dawn and evening can be then best used. Smoke from forest and moss fires, of course could black out the whole country. Another problem was the fact that the ray was never more than a few hundred feet above ground for all of its 5 to 20 miles [8–32 km] of length and hence suffered the full effects of heat haze and refraction. The sun almost invariably had to be from the observer to the target, a factor that made horizon closures difficult. A test with night lights, using Coleman gas lanterns, proved effective, although trouble was experienced from ground fogs.... Reading angles on the job was difficult even for our most experienced men, who although familiar with the normal observing hazards, were sometimes baffled by the difficulties of judging distance, of finding direction or locating landmarks, and by the tricks of the critical light.

Despite the difficulties and some of the trial and error procedures, the 1953 field season produced a successful start to the difficult surveying landscape of northeastern BC.

In his government report Andrews described another significant project in 1953:

The right-of-way survey for the Trans-Mountain Oil Pipe Line is another important project in the category of legal surveys, being now almost completed after nearly two years of intensive effort by all concerned. The survey follows the long pipeline through British Columbia for some 500 miles [800 km] from Yellowhead to Burnaby. Done at company expense,

Station Bubbles stands
24 metres tall to survey
above the trees.

under the direction of British Columbia land surveyors, it also involved
a great deal of work in this Branch at Victoria and in the offices of Land
Registrars at Kamloops and New Westminster. Being the first survey of
its kind in British Columbia's history (and we hope not the last), and the
necessity on the company's part to have the easements registered prior
to construction, engendered a great many problems of procedure, interpretation, and clarification of statutory requirements, drawing up suitable
instructions for survey, checking and approving plans, and restoration of
survey monuments disturbed by construction....

A by-product derived from experience with the Trans-Mountain Oil
Pipe Line survey has been a set of surveyor general's instructions for survey of rights-of-way (other than highways), which was issued under the
date of July 15, 1953, and which should clarify procedures to all who may
be concerned with these matters in the future.

1954

The survey in northeastern BC continued in 1954:

The Topographic Division experienced its second year in the field ex-
clusively devoted to extending the unique triangulation network over
the muskeg lands of northeast British Columbia. Thanks to experience
and skills gained in tower construction during 1953, to a more liberal
helicopter contract, and to the low-cost services of our own transport
aircraft, a record area of new but difficult country was transformed from

Moving supplies by helicopter.

the category of the unknown into that of the known. It is anticipated that the 1955 season will complete this program of triangulation controlling 25,000 square miles [65,000 km²] of hitherto unsurveyed potential oil and gas reserves.

Alf Slocomb, head of the Topographic Division, provided more information about this unusual survey:

With a definite objective of the 59th parallel of latitude and a possible objective of the 60th parallel set at the beginning of the season, excellent progress was made until the Fontas River section was reached. This area of low relief proved very difficult, necessitating towers of well over 100 feet [30 m] in height to carry the work forward. Despite this the first objective was easily attained, plus approximately one-third of the second....

High above the Parliament Buildings at Victoria is a statue of Capt. George Vancouver. It is 165 feet [50 m] from the ground level to the top of Captain Vancouver's outstretched hand. The highest triangulation tower built this summer was at station "Strip" where the recorded measurement to the top of the flag was 156 feet. The building of such a tower with materials found at the site plus great quantities of spikes and wire is a personal triumph that requires a fine sense of proportion, stamina, hard work, and the nerve to climb and work at such heights. The only equipment available to the builders are: axes, hammers, saws, block and tackle, and muscles. No plans or blueprints can be supplied, as each tower site presents its own particular problem and the building materials vary considerably. The builders also have to contend with the usual nuisances – mosquitos, black flies, etc., inquisitive bears that can be very persistent, and the local thunder storms which are characteristic of this section of the province.

Air transport is the only economic and rapid means of access to this area of shallow lakes and muskeg. Transporting men, equipment and supplies for a fifty-man field party is a major undertaking. We were fortunate to have had the use of the Air Division's Beaver aircraft FHF, equipped with pontoons. Able to fly on many of the small lakes, it was the work horse of the operation, logging 567 hours flying time. Every pound of freight came in by the Beaver, and all the preliminary reconnaissance trips were squeezed into a long tight schedule. Certainly the objective obtained would have been impossible without it, and to charter for a like number of hours would have been far beyond our financial resources.

We chartered a Bell helicopter from Okanagan Helicopters Limited which logged 633 flying hours. It was used for transporting men and equipment to the individual stations and for station reconnaissance where its ability to hover at tree heights was particularly useful. On three separate occasions a second machine was hired to take care of the peak periods, for a total of 137 additional flying hours.

In another survey in northern BC in 1954, the right-of-way survey of the Alaska Highway in British Columbia, a project that had taken several years, was completed. This survey was made at a high level of precision so that future surveys in the area could be connected with it.

1955

The northeastern BC survey was finished on schedule in 1955. Ed Bassett, who became the Deputy Minister of Lands in 1954, observed that "the value of this control to British Columbia is immense, for it means that legally correct recording of oil, natural gas and associated locations can be established from the Peace River district to the northern boundary of British Columbia." Alf Slocomb trumpeted the completion of the project:

The seemingly impossible has become a reality.... Like most things apparently impossible there is usually an answer if you have the wit to see and the will to do. Tower building was the magic solution, the helicopter was the magic carpet, and a group of agile, fearless young men, most certainly akin to squirrels were the magicians. Twenty-six thousand square miles [67,000 km²] of controlled territory was the final score, and one to be proud of.

Ernie McMinn wrote about some of the details of the final year of this project:

Spring, and the survey party came to Fort Nelson on the second day of June to find the great river already humming with activity. Four float planes, using the muddy stream as a base were servicing oil exploration

Gathering supplies at Fort Nelson for the field season.

crews and an Army survey party. The 60th parallel survey crew under W.N. Papove, BCLS, DLS, were caulking and tarring their two freight boats preparatory to their 200-mile [320-km] access trip to the Beaver River. River barges were loading supplies for Aklavik. Our six truck convoy unloaded ten tons of equipment at the river bank and the two three-ton trucks left immediately for Dawson Creek on the first of many 600-mile trips for food and gasoline.

Under the spruce trees near Fort Nelson a supply and transit camp of eleven tents was erected. A road was bulldozed down to the river's edge, where a serviceable log float and refuelling facilities were completed just as our yellow Beaver FHF came in on schedule from Victoria.

Following the same pattern of work as the previous two years, the surveyors reached the northeastern corner of British Columbia:

Here, in a little muskeg, where the Boundary Commission pipe post indicates the intersection of 120° west longitude and 60° latitude, a 57-foot tower was built and named "Artieboy" after our good companion of two seasons, Arthur Coles, who died in the crash of his helicopter this summer in Ontario.

After reaching the objective of our three seasons' work, we turned westward toward the Liard River. Observing parties commenced a rear guard action, completing the angle reading as we moved along. At this time progress was threatened by a forest fire which had been burning for two weeks along the Alberta boundary and which now paid us a visit, loading the atmosphere with smoke and reaching 30 miles [50 km] into our area. It burned one side of the 90-foot tower "Peggo"....

From June, when the sun rose and set in the same northern sky, the triangulation network was woven across over 8,000 square miles [20,000 km²] of potentially valuable muskeg lands, until we reached and crossed

Surveying from one of
the towers.

the Simpson Trail in August – August with its lengthening darkness and
nightly display of the aurora borealis....

By the first week of September, student helpers were leaving to return
to school, but the tower builders were at their last stations.... On the other
hand, the instrument observing work was lagging, mostly because of the
loss of experienced men. Two towers on exposed ridges, Petitot and Dilly,
blew over in the August gales and were rebuilt. The bright, hot sunlight of
June and the eighteen hours of daylight were gone; we finished the angle
reading in autumn with its pale, hazy sunlight and its eleven hour day....
Eventually the observing program was completed satisfactorily, if not per-
fectly, in the light snow storms of October.

The crew made five ties with the Alberta boundary survey and ten with
the 60th parallel. "The key reference point of the petroleum and natural gas
permits in the western section – namely the old Hudson's Bay Company
post at Nelson Forks – was also fixed." (This was a survey location established by Milligan in 1914.) The men also made ties to the Alaska Highway
right-of-way surveys. Andrews wrote: "Our 25,000-square-mile reservoir of
potential petroleum and natural gas has at last a fence around it, so to speak,
of perimeter surveys."

The men in the field had done the work that made this project possible,
but Andrews had provided the leadership and organized the Surveys and
Mapping Branch to utilize all facets of the expertise of its personnel and
various divisions to make the northeastern survey a success.

McMinn summarized the scale of this three-year surveying program in a
remote area of the province:

In three years 26,000 square miles [67,000 km²] of country have been
covered, using 184 stations, most of which were towers. The total combined height of these towers is 11,400 feet [3,575 m]; twenty-four stations

were over 100 feet [30 m]. Flying time on the helicopters was 1,588 hours; on the Beaver 1,042 hours. Nearly a million pounds of freight, including 40,000 gallons of aviation fuel, was transported by truck, Beaver, and helicopter. Ten tons of spikes, 4,000 yards of signal cotton, and 7 tons of cement were used. The triangulation network, which involved some 212 stations and the reading of 26,000 angles, included the Geodetic stations from Fort St John to Fort Nelson and ties to the Alberta boundary, 60th parallel monuments, the benchmarks on the Alaska Highway, Beatton Road, and Simpson Trail, and to all location posts established under the Petroleum and Natural Gas Act.

To facilitate the coordination between the surveying and the oil and gas tenures the province adopted a grid system under the *Petroleum and Natural Gas Act, 1954*. Andrews described it:

> The grid system provides a nice basis for specifying size, shape, location, and precise areas without risk of excess or deficiency on the ground between adjacent tenures. The job of locating the features of the grid system, such as well sites, lease corners, etc. on the ground is a straightforward survey problem, fully capable of execution by qualified surveyors working under official regulations and instructions.

Although the labour in the field had been completed, there was now a large amount of office work to be done. Bert Ralfs, the assistant director of the Surveys and Mapping Branch, reported:

> Our heavy program in the last three years has produced a tremendous crop of triangulation and other control.... When the oil activity began there was no basic geodetic control throughout the area, much less such control being on a final datum. All of the calculations accordingly have been based on what is known as the "Prince George non-closure final" on which datum the positions are preliminary only and differ in varying amounts up to about 90 feet [27 m] from the final datum positions.... In effect we are committed at present, and until the pressure eases, to maintaining two sets of coordinates.

Once all the data from the 1953 to 1955 survey were calculated the surveys done before that date would be converted to the new network. Fortunately the federal government was helping by creating maps from the data. In the meantime, "the available field survey information – coordinates, azimuths, elevations and photo identification – has been supplied to many oil exploration companies and commercial mapping organizations."

Surveys and Mapping Branch also worked on other significant projects in 1955. The deputy minister noted that "the last gap of 66 miles [106 km] has been completed in that section of the northern boundary of British Columbia east of Teslin Lake." Andrews wrote that the survey of the 60th parallel was now completed to the western part of Yukon, leaving only two remaining sections: a resurvey of 260 kilometres to the Tatshenshini River that was originally done shortly after the Klondike Gold rush, after which "a decision may be in order whether to tackle the furthermost westerly 80 miles [130 km], beyond the Tatshenshini River, which has never been surveyed, due to extreme inaccessibility, and to the prevalence of high ice-burdened mountains in that region." In the same year the Alberta and British Columbia governments passed legislation to accept the final surveys marking their border, followed by legislation from the federal government. As BC's representative, Andrews contributed to these details and the final Boundary Commission's report.

As he was completing work on this commission Andrews was appointed to the Fraser River Board the same year. This was an intergovernmental committee that existed for several years. It examined management of the river, particularly in relationship to flood control and hydroelectric power.

Andrews reported that in 1955 the Air Surveys Division

> contributed a record total of basic data toward fundamental investigations in a variety of resource development projects. Its air photo production was for such major practical matters as forest public working circles, more intense forest inventory survey, and new forest road locations. Greatly increased multiplex mapping services were extended to many agencies, such as the British Columbia Power Commission, the Water Rights Branch, the Fraser River Board and the Forest Engineering Division.... [The] preparation of major and complicated easements required much and detailed attention. Right-of-way for the Westcoast Transmission pipeline and for both the British Columbia Power Commission and the British Columbia Electric Company are cases in point.

Water projects, particularly those related to hydroelectricity, remained important. The Surveys and Mapping Branch was involved with studies by the Fraser River Board; a research program for the Columbia River basin "to obtain facts which would assist the British Columbia government to formulate policy concerning the future use of these waters"; and an inspection of the Atlin area following the Northwest Power Industries proposal to "divert and use the waters of the Yukon River, Teslin Lake, and Taku River for a proposed hydroelectric power development of enormous potential".

1956

The economic development of British Columbia continued unabated in 1956. In his report for the year, Deputy Minister of Lands Ed Bassett wrote:

> The volume and variety of land and water surveys, mapping, alienations and development have increased over the previous record year of 1955.

The inference is that British Columbians today are living in an era of continuing general prosperity, increasingly higher levels of living standards, and rapidly mounting primary and secondary industrial activity with attendant and multiplying services. British Columbia is in the happy position of being both a land of promise for the future and a land of economic and cultural dividends in the present.

Bassett listed a dizzying number of projects that involved the Surveys and Mapping Branch, among them "the granting of easements of rights of way for industrial enterprises. Those granted to the Westcoast Transmission Company for its natural gas pipeline from the BC Peace River District to Vancouver are perhaps outstanding in the year's accomplishment." The branch also surveyed easements for West Kootenay Power, the British Columbia Power Commission and the British Columbia Electric Company. Bassett noted that "the Lands Branch boils over with the discharge of its many other land and water alienation responsibilities" along with the cooperative work it undertook with many other provincial and federal government agencies. He also stated:

Substantial progress was made in 1956 in the field of water management and planning and special investigations of great significance were started during the year. Among these special scientific investigations were the commencement of the preliminary studies of the Peace River and the Dease River with the purpose of evaluating within British Columbia, the hydroelectric resources of these streams.... [Work also started] to place together all the basic data on the Columbia River system in order to prepare a report for the government of British Columbia concerning the best development of this river from the point of view of the interests of the province.

Andrews described some of the topographic surveys his organization did:

Having butted up against the province's north boundary in 1955 with its triangulation program in northeast British Columbia, this year on the rebound, the Topographic Division carried out similar work in the unsurveyed wedge of country south of the Peace River Block ... where locations and exploration under the *Petroleum and Natural Gas Act* have been active and in need of survey control. Here again the well-developed technique of tower building was used in the low timbered basins of the Kiskatinaw and Beaverlodge rivers, but toward the south and west part of the area the triangulation network ascended the foothills to the outlying ranges of the Rockies, where reversion to bare mountaintop stations was a welcome change after four seasons of tower building in the lowlands.

The Surveys and Mapping Branch conducted two surveys for private companies. One was for the British Columbia Power Corporation in the Chilko and Taseko lakes area.

In the foothills of the Rockies the oil and gas triangulation survey returned to the ground.

The other project, comparatively small but spectacular, was a triangulation across the Salmon and related glaciers from the Granduc mine to a point in the vicinity of Stewart at the head of Portland Canal. This was to provide azimuth and levels between the two ends of a proposed 18-mile [29-km] tunnel.... The Granduc Mining Company provided transportation facilities by aeroplane and snow tractor on the glacier. The company also provided cabins on the glacier for the use of the survey crew, as well as living amenities at the mine for the work at that end....

Both projects ... witness the fact that for major survey control, which is indispensable, our government topographic surveyors are acknowledged as eminent specialists.

Andrews received an unusual assignment that year, to supervise the direction of subdivision surveys of the Doukhobor lands according to recommendations by the Royal Commission currently inquiring into this matter. The scheme of subdivision was too large to be done by staff surveyors, who were fully committed on routine work, but the task of preliminary subdivision design, preparation of instructions, broad supervision, checking and approval of the final plans and accounts were all handled by the Legal Division staff. The actual job was assigned to a firm of private surveyors.

The British Columbia–Yukon and Northwest Territories Boundary Commission decided to continue surveying the 60th parallel west of Teslin Lake. This line had originally been surveyed between 1899 and 1901, so the Commission's instructions to Art Swannell stated:

This portion of the boundary, extending west from Teslin Lake to Takhini River ... should be re-surveyed, but not re-located, to modernize the standards of accuracy, monumentation and line cutting. The monuments were to be refurbished or re-established using modern rock or soil posts accompanied by pits and mounds and if possible, bearing trees.

The commission also directed Swannell to establish a "triangulation network, straddling the line if possible, so as to be able to tie in a number of the monuments along it." This would enable future surveys in the area to be tied into the 60th parallel.

In his report Swannell explained that conditions on the ground dictated a change in the surveying. "Originally it was planned to commence operations at Teslin Lake, but because of the unusually late spring, ice conditions there prohibited us from getting to the boundary line, so a start was made on actual field operations on June 1st at Atlin Lake." Once he and his crew, which included up to 18 people, reached Teslin Lake they returned to Atlin and proceeded west for 113 kilometres beyond Teslin Lake. Swannell wrote: "The last good weather was experienced on September 15. On this day, a most beautiful one, the Boundary Commissioner, Mr G.S. Andrews, Mr New [Swannell's assistant] and myself took a reconnaissance flight over the remaining portion of the boundary to be retraced. A beautiful day and flight, but a very rugged country." Swannell's party used horses but he noted: "This could well be one of the last sizeable survey operations to use horse transport, especially if the performance and supply of helicopters continue to improve as they have done in recent years."

The Air Surveys Division processed its one-millionth aerial photograph and the air-photo library continued to be used by provincial and federal government agencies as well as the public. "There has been a noticeable increase in use of its facilities by research scientists and graduate students."

According to Bassett, "it is no wonder then, that the pace or tempo of activity within the Surveys and Mapping Service during 1956 has been, to say the least, brisk and humming. All divisions of the Branch – Air, Geographic, Legal and Topographic – have been and are, under great pressure of work to meet the unprecedented demand for surveys and mapping service."

Despite the department's continued success, Andrews had some cautious observations regarding all the activity and work:

> The problems of manpower ... are accentuated in government service
> where remuneration has been traditionally non-competitive with industry.
> This has the effect of making government service a training agency for
> industry, since, only too frequently, as soon as a raw recruit has reached the
> point where the investment in his training begins to pay off, his compe-
> tence and experience command higher remuneration outside.
>
> On the other hand, the expanding economy of the Province has stimu-
> lated an unprecedented demand on government for surveys and mapping.
> These circumstances are forcing a metamorphoses upon us in the direc-

tion of automation, and it is propitious that a number of recent technical developments give promise of progress in that direction.

He also commented:

> For the first time since World War II we must admit that the air photography taken by this branch, while still excellent, is no longer unexcelled. Our air photographs, unsurpassed for a decade, are now obsolescent, and our multiplex plotting equipment is classified as third order in precision. The branch has long been recognized as a pioneer especially in air photogrammetry and in the use of helicopters. It should continue its leadership in scientific progress.

In his report Andrews discussed the prospect of acquiring some new, more modern equipment, concluding with these thoughts:

> Such technological developments will go a long way toward meeting the demand for surveys without unduly increasing competition for manpower. It is a fact that such equipment calls for considerable capital investment, the carrying charges on which will tend to keep mapping costs per square mile at about the same level as the past, but our capacity for work will be nevertheless enhanced....
>
> Surveying and mapping is being forced into big business. The day of the simple theodolite, the chain, and the packhorse is over.

1957

> The volume and variety of activities by the British Columbia Lands Service – land and water surveys, mapping, and alienations from the Crown – form a strong indication of economic conditions in the province. In summing up the accomplishments of the Lands Service in 1957 it is a pleasure to record that the great interest in land and water acquisition, development, and use has continued unabated and has even exceeded last year's record high.

So began the 1957 report from Deputy Minister of Lands Ed Bassett.

More pragmatically, Andrews reported a new branch record, "the charter of three helicopters for the full four-month season". The Surveys and Mapping Branch paid for one helicopter that was used on several surveying projects. The second was financed by the Water Branch for their studies of the Parsnip and upper Peace rivers, while the third was funded by the Boundary Commission. "This extensive experience with helicopter operation confirmed our past findings that the calibre of helicopter personnel (pilots and engineers) assigned to survey operations in mountain areas is of greater importance as a factor of efficient operation than the actual tariff paid for the machines in dollars."

The changing technology that Andrews discussed in his report for the previous year was demonstrated in 1957:

> Anticipating future use of electronic computing machines for high-speed processing of the rather involved mathematical adjustment of survey control data, the Geographic Division arranged for the supervisor of its Computing Section to participate in a special night course offered at the University of British Columbia during the winter months of 1957–58....
>
> [The branch also purchased] a new electronic device, called the tellurometer for measurement of distances ranging between 1,000 feet [300 m] and 25 miles [40 km].... One of these outfits, ordered as soon as funds were authorized for the current fiscal year, was received toward the end of the field season, too late for operational use, but in subsequent tests and training it has confirmed its remarkable accuracy and simplicity of operation. Significant improvements in the flexibility, accuracy and economy of our future control surveys are anticipated by the use of this equipment.

The Surveys and Mapping Branch was the first government organization in Canada to purchase a tellurometer.

Andrews commented on an issue that had a possible effect on economic growth:

> The unavoidable rise in costs of legal surveys of Crown lands, especially in remote areas such as in parts of the Peace River District, is a matter of much concern. Even when this work is restricted to concentrated groups of parcels, the cost per acre for legal surveys into district lots pertaining to applications for Crown lands is often higher than the statutory sale price per acre under the *Land Act* and in excess of the presently used surcharge for survey per acre.

The government would either have to subsidise the cost to maintain the strong public demand for land, particularly in the northern part of the province, or would have to increase fees charged to the purchasers of the land and risk a decrease in sales because much of the land had marginal economic value.

An even more serious issue arose during 1957:

> A major problem in the operations and finances of the Air Division arose this year due to an abrupt change in the specifications for air photos and map compilation required for a new phase of the provincial forest inventory. Heretofore the basic program of air photography and interim mapping, at a scale of one half mile per inch, has been financed to a large extent by federal-provincial moneys available for the national forest inventory under the *Canada Forest Act*, and has constituted the main activity of the Air Division, both in the field (air photography) and in the office (map compilation). The versatile usefulness of this program, however, was such that its resultant air photos, some 200,000 now in the Provincial Air Photo

Library, and maps, some 1,500 reproducible fair-drawn master sheets on file in the department, have been of primary importance to all other government departments and private interests concerned with the opening up of the country and the wise use of natural resources.

Until 1957 most of the aerial photography the Surveys and Mapping Branch had done was for the Forest Service and had been financed by them, but other departments had been able to utilize the photos and maps produced from them because they used the standard scale. The new photographs that forestry wanted were not "adaptable to other versatile uses, such as detailed contour mapping" and would not cover any territory north of 57° latitude. Andrews reluctantly pointed out:

The coincidence of the air survey requirements of the forest inventory program with general purpose basic mapping of the northern third of the province, with its very important financial contribution, is obviously now at an end. To carry on the program into the north, additional money for both field and office operations will be required.

He did make a case for continuing the mapping of the far northern part of the province:

To offset the loss of the forest inventory refund for continuation of the basic interim air mapping in the north ... an additional annual allotment of moneys will be required. The total would be much less than the cost of one mile of first class highway construction in average British Columbia terrain.

Art Swannell continued the BC–Yukon boundary survey westward. The Boundary Commission report stated that Swannell had been

instructed to continue the retracement of the original survey of the boundary by Saint Cyr between Teslin Lake and the Takhini River (Mon. 118) and to survey the existing unsurveyed gap of nine miles between Monument 118 and Monument 119 at Hendon River.... It was decided that triangulation methods would be used to determine the distances between the monuments and also, that a helicopter would be employed for field transport in this rugged area....

Before leaving Victoria to begin the field operations, Swannell compiled a preliminary map from aerial photographs and used it to plan a suitable triangulation scheme. This was very successful and simplified the field work considerably.

The original boundary survey had been done over 50 years ago in rugged country and under difficult conditions, but surprisingly, Swannell was able to find all but one of the monuments. This one was in a valley and had probably been washed out in a flood. Swannell successfully completed the commission's objectives for the season.

The area north of Prince George, particularly around Tudyah Lake, was

Tudyah Lake camp, featuring Melville's water wheel. The Beaver float plane used by the surveyors is moored at the shore.

a busy location for surveying during the summer of 1957. Frank Speed and his crew used a boat and helicopter to conduct surveys in the Rocky Mountain Trench along the Parsnip and Peace rivers. These were a continuation of surveys for the Water Rights Branch undertaken in the previous few years in connection with proposed dam sites on the Peace River. In the nearby Rocky Mountains Duff Wight established survey control for several developments occurring there. He set up a supply and base camp at Tudyah Lake for his and Speed's crews.

A Summer Sojourn with a Stenorette

Andrews' 1957 summer trip to visit the surveyors in the field was one of the few occasions when he went by himself. To record his trip he took a stenorette. This was a small dictation machine with a hard plastic body that was a predecessor and cousin of the portable audio cassette player. The tape, which could take 30 minutes of recording, was on an open reel that was slightly larger than an audiocassette. Although primarily designed for office work, the stenorette had a connection that would enable it to use a car battery for power, giving it portability.

Andrews left Victoria on July 31 in a Willys four-wheel drive station wagon. On the way, he visited Jesmond and the area where he taught school at Big Bar. On August 2 he started using the stenorette:

> Testing. Just installed the equipment at Marguerite. The converter
> seems to make a humming noise but I suppose that is normal.... Last
> night stopped at Lac La Hache, had a nice visit with Art Barber. Met his
> crew – Rudolph, Ted and Tom. Art thinks it will take them a bit longer
> than estimated to do that job on account of the fact that the vertical angle

shots laid down don't appear to be too practical on account of timber interference. On leaving Lac La Hache this morning, I stopped in to Gilbert Forbes' place, father of Tom Forbes, who is with Duff Wight on the topographic survey. Forbes worked for Frank Swannell on the Nechako Survey in the 1910 era and remembers R.P. Bishop on that same party.

By mid afternoon Andrews was "about 15 miles out of Prince George. Just on to the pavement, thank God. Weather appears to be clearing somewhat. Note: should have a dust cover for the stenorette for travelling." He had supper at McLeod Lake north of Prince George and then "called in to see McIntyre at McLeod Lake and Bill

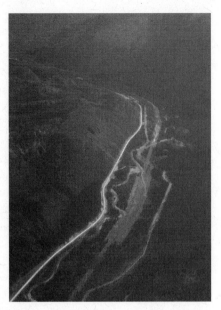

Flying over Pine Pass. The Hart Highway is the pale line on the left.

McPhee at the Forest Ranger Station." Andrews had first met McIntyre on his 1924 pack trip through the area, and McPhee was a friend from his years in the Forest Service. Stopping for the evening at Tudyah Lake, Andrews "found nobody at the survey camp.... Have a room at Melville's Lodge on the speculation that Lamont may come in in the morning or someone may show up.... Melville's place at Tudyah Lake is a trailer metropolis. There must be more than 50 huge trailers accommodating the various workers on the Westcoast Transmission Pipe Line job."

The next morning Andrews talked via radio to several surveyors who were working in the region, checking on the weather in their area, their surveying progress, and making arrangements to see some of the work. Late in the afternoon he gave a summary of the day:

4:30 pm. Just leaving Parsnip Bridge for Azouzetta Lake. We had a good flight in the Beaver. First, from Tudyah to the Nation River camp and then on to Finlay Forks where Speed is establishing an advance camp, later to his main camp, and then we took off there after lunch in the Beaver and took a circuit up the Ospika River and came back across the Peace at about the Ne-parle-pas Rapids, and then followed up the Clearwater River on the south side of the Peace watershed, and then over to the Pine and Pine Pass and down the Misinchinka, Back to Tudyah, getting in about 4 o'clock.

The helicopter was operating from Nation River and had been out one or two trips this morning. They were not able to make one or two landings on account of the high wind, but were able to get down to alternative stations.... Had a nice visit at Finlay Forks with Roy McDougall and his wife, Marjorie, and their granddaughter little Marj.... Crossing over Pine Pass on the flight back today, it was interesting to see the road and the PGE right-of-way slashed pretty well through and some grading done, and then of course the gas pipeline meandering around wherever it could get good going, after the road and railroad having taken first priority on the best going. Some beautiful alpine country on the hills immediately north of the Peace River toward the west end between the Peace and the Ospika and more of it just above Pine Pass to the north and west of Azouzetta Lake. Finlay Forks looked very much the same as it did in 1939.

Andrews camped that evening at Azouzetta Lake, and spent time looking at some of the places around the lake where he had been in 1924. The next day he drove on to Dawson Creek:

Pine Pass is a scene of activity with the PGE construction and the pipeline just completed through there. Also the radio telephone system for the PGE. Building relay stations through. An impressive contrast to my first trip through here over 30 years ago with the horses. Wight's operation got a bit of work done yesterday but were troubled with high winds and working in pretty high country. He has no alternative low stations to do when he can't work on the high stations on account of wind. The country itself looks much the same as it did, however, in the old days.

Andrews arrived at Dawson Creek in the early afternoon and visited surveyor Jim Mackenzie at his house. "He's batching there. Wife and baby down at the coast.... Jim confines his activities pretty well to Dawson Creek now." Then he continued on to Fort St John. He couldn't find surveyor Duncan Cran so he moved on to Rose Prairie to meet Jim Young, his friend from teaching at Kelly Lake.

His next recording on the stenorette was on the evening of August 8 where he highlighted the events of the previous four days. After staying overnight at Young's place the two men drove to Hudson's Hope to visit some people. In the early evening Andrews visited Duncan Cran before taking a CP Airlines flight to Whitehorse, where he arrived shortly after 10 pm. The next morning he took the train to Carcross:

I had just time to have some coffee with Mrs Simmons and we heard Peterson coming in by plane. Mr Aubrey Simmons drove me over to the landing and Peterson and I took off for Swannell's camp at Primrose Lake. The weather was pretty solid over the higher peaks, but fairly good visibility below that, so there was no difficulty in navigating to Swannell's camp.

Atlin from Harper Reed's rooftop.

We got to Swannell's in time for an early lunch and had a good session with Art. Found everything progressing quite well, in spite of indifferent weather, and so I came back. There wasn't much point in staying there, waiting indefinitely for the weather to clear up, so I returned with Peterson after lunch.

Just as we took off from Primrose Lake, Peterson told me that he was going down to Telegraph Creek and to Cold Fish Lake the next day, and gave me an invitation to come along with him if I could spare the time. So I decided it was a good opportunity, so we didn't bother going back to Carcross but went straight on to Atlin where we landed about 2 o'clock in the afternoon. Got a room at the Kootenay Hotel there and then looked up Harper Reed. Harper was busy jacking up his house, preparatory to moving it up to a new lot nearer the middle of town that he owns, where he hopes to stay next winter. Harper Reed closed down operations immediately and took me for a diplomatic tour of Atlin.

Andrews and Reed spent the afternoon visiting several people and had supper at a coffee shop.

Wednesday morning the 7th, I got up at 5:15 am and went down to Fred Peterson's place with my stuff where he had a breakfast ready. Mrs Peterson had been up, got it fixed up, and we took off about ten minutes past seven. We got to Telegraph Creek about 10 in the morning, Yukon Standard time, and flew pretty well the route of the old Telegraph Trail,

Tommy Walker's camp at Cold Fish Lake.

almost right on it all the way. At Telegraph Creek we dropped quite a sizeable load of mail, about 15 sacks, and odds and ends.

Peterson had expected to find four ladies there, including Mrs Dr Norman McKenzie, who were to go into Cold Fish Lake. They were coming up on the river boat from Wrangell and were supposed to be at Telegraph Creek when we arrived. However, they were not there and there was no knowledge of when the boat, the *Judith Ann*, would be in. Meanwhile we had a nice visit with Steele Hyland and Lou. Met the local Mounted Policeman, Mr Brand.

Rather than wait around Telegraph Creek wondering, Herman decided we should take off and fly down the river to see where the boat was. This we did. Got off about half past ten and kept flying along the river farther and farther with no sight of the boat. Finally, we located it just coming around the bend below what they call the Clearwater River coming in from the Northwest. Her speed appeared to be almost backwards, it was so slow compared to our means of travelling. The junction at this point of the two valleys gave lots of room for manoeuvring the aircraft and we located a fairly good run of the river above the *Judith Ann* for landing, so we circled the ship low and then landed at this place and succeeded in tying up on the right hand bank of the river and made fast to a big pile of driftwood there. The water was nice and deep on the convex side of the stream and it was a pretty good spot.

The *Judith Ann* sent a small boat over to us with a kicker, and we told the lad that we wanted the four ladies if possible and their baggage to be transferred to the aircraft.... In due course the four ladies were brought ashore complete with their baggage and then they were carefully put on the plane and we took off, the whole affair going very nicely and nobody got wet. By pre-arrangement I was not introduced to the ladies excepting as "Gerry the Swamper".

In a few minutes we were at Telegraph Creek again and the ladies all got out there and looked over the town with Mrs Hyland while we loaded the freight and other supplies for Cold Fish Lake. It was a bit after twelve noon when we finally took off with quite a heavy load.... The flight from Telegraph Creek to Cold Fish Lake took just 55 minutes, right on normal time, and we made a nice landing.... I was duly introduced at this stage to the lady guests as Gerry Andrews, surveyor general of British Columbia. The Walkers exhibited surprise and very genuine delight in my arriving at Cold Fish Lake unexpectedly.

In the afternoon, while the plane was delivering supplies to another location, Andrews walked down to the end of the lake.

The plane came back about 4 o'clock and apparently they had forgotten some important supplies for the Buckinghorse Lake camp, so we had a quick supper and put on all the mail to go outside and these overlooked supplies, and Herman Peterson and myself took off for Buckinghorse Lake across the Spatsizi River. We got in there in less than half an hour and delivered the supplies.... We took off again about 5 o'clock or a bit later, and flew directly across country to Telegraph Creek and again there was a little exchange of mail and other odds and ends. Taking on a bit of gas. Finally got off Telegraph Creek about 7:30 in the evening. Climbed straight up and headed back toward Atlin on our normal course. It was a beautiful flight back, odd rain showers scattered all over the country. Very lovely illumination effects from the declining sun.... Ahead of us in the direction of the sunset or the twilight there was a huge open area in the sky which apparently was over Atlin Lake. We finally made a landing about 9:15 Yukon Standard Time. It was a most wonderful day altogether.

Following the flight Andrews took a taxi to Whitehorse, arriving there after midnight where he got a room at the Whitehorse Inn. The next morning Andrews flew back to Fort St John:

We hit some very high convection storms and had one of the roughest trips for a while that I have experienced for a long time, but we landed on time in St John.... Had a bit of trouble getting the car onto a greasing rack and then they found that the muffler had come adrift underneath and was fouling the hand brake cable, so they fixed that. Meantime, I met George Murray, Mrs Murray and Georgina, and had supper with Alwin Holland

and Mr Bouffioux and Duncan Cran [surveyors] and young Billy Robertson, Duncan's 14-year-old helper.

At 10:50 that evening Andrews recorded that he had arrived at Mile 147 on the Alaska Highway where he was stopping for the night. His next entry was at 9:20 the following morning:

> Just leaving Mile 147 on the Alaska Highway, heading for Dawson Creek and Hart Highway. Have just been in the morning already to see Tony McLaughlin on the Halfway River. We took off [in a helicopter] for the Halfway about 8 o'clock and were there by ten past eight or a little more. Found Tony camped on the west side of the river. He had just moved yesterday to this new location from the other side of the river. Apparently they had a heavy downpour of rain yesterday during their move and the camp looked pretty wet this morning.... The Halfway valley is a pretty attractive looking bit of country. Also a subsidiary creek, Cameron Creek, a nice incidence of open meadow land here and there, and some rather nice looking pioneer ranches. McLaughlin has surveyed in several Crown lots vacant over good looking pieces of the valley.

Andrews drove back to Fort St John, then south to Dawson Creek where he spent a few hours. "3:30 pm. Just leaving Dawson Creek, heading for Pine Pass. Still raining steadily and the roads are pretty sloppy. Had a cup of coffee with Jimmy Mackenzie at his house and checked with Mary Young at Pappy's Service Station, so now I am all clear for the Peace River country." In the early evening he arrived at Duff Wight's camp in Pine Pass.

On August 10 he left camp at Azouzetta Lake. "Very heavy rain all night and continuing.... Memo: should get new photographs of this Pine Pass route when the construction work is finished to show the access roads from the highway to the railroad grade at different places. Some of those might be quite useful for developing park areas and camp sites." After a stop at Tudyah Lake he continued to Prince George.

> Got a call through to Dick Leak [a surveyor] at Vanderhoof telling him I intended to come up the next day. Phoned Mr Burden [another surveyor] who invited me up to the house for the evening after supper and had a very pleasant visit there, and as a result of his conversation about the Vanderhoof road, I decided to go up there by train on Sunday.... It was still raining and wet, so left Prince George at about 7 am Daylight Time on the CNR. I got to Vanderhoof about half past 10 or 11 and took a taxi out to Dick's camp. Had about two hours with him there, and then caught the train back to Prince George.

After having supper with Burden, several people came over to visit and Andrews didn't get back to the hotel until about 2 am. "Found a message from Don Whyte, BCLS, who wanted to see me so it was too late to get him at that time of the night. This morning, the 12th, found Don Whyte at Keller House waiting to have breakfast, so we had breakfast together and a bit of a chat." Later in the morning Andrews "stopped at Stone Creek, south of Prince George, 20 miles, for a final talk to Don Whyte, BCLS. He explained that they were having a little difficulty with some of the lot structures in this area. Made a note of that in my notebook."

Around 3:30 that afternoon:

> I've just been pulled out of a mud hole on the Cinema Hill by a wrecking truck from Johnson Motors, in Quesnel. Coming up the Cinema Hill at a left hand turn, I was passing a truck and he crowded me a little bit. My left wheel got into the soft shoulder and the car nearly rolled down into a steep gully. Very luckily, it balanced in a critical position. I was able to crawl out. However, I wasn't able to do much around the vehicle for fear that it would go at any minute. I got the truck to try to pull me out with my own winch cable, but it snapped of course, so I used it to make fast to a tree on the other side of the road in such a way that the cable lay flat on the road not to impede traffic.... A passing car offered to take me in but I asked him to report to the Esso station to send a wrecker out. I waited about I guess an hour and a quarter for the wrecker, and he didn't show up, so I picked up a ride in toward town but met the wrecker a couple of miles down the road and returned to the scene of the accident with him. The wrecker man hitched one line to the right hand aft end of the jeep around the frame and he put the other line from the derrick over to a tree on the right hand side of the road and just hauled her out with a winch very slowly. Apparently no damage was done to the Willys.

> I have now reached the pavement approaching Quesnel. It's 3:45 pm and the little car seems to be perfectly normal, except I think the hand brake may have gotten a bit of a pull.

Andrews stayed at Lac La Hache that night and

> had a pleasant evening with Art Barber [who was doing a small project requested by the Water Rights Branch] and his wife and his crew and children. He is breaking camp today and Rudolph Helmet is going north to join the Topographic at Tudyah and the other two boys and Art are going south. Tom and Ted are going to go to Ken Bridge's show and Art is going to Victoria prior to going north himself to the Topographic.

Later in the day Andrews stopped at Lillooet to visit retired surveyor Geoffrey Downton who was in bed recovering from an operation. "He was very glad to see me and we had a good chat." After spending the night in Lytton, Andrews drove to Vancouver on August 14, where he took the night boat to Victoria. Andrews' last entry on the stenorette was on August 15. "2:55 pm. Have arrived in Victoria and checked in at the house and unpacked, and now heading for the office.... The total mileage for the trip was 2,343 miles [3,771 km]."

Andrews' trip in 1957 was typical of his summer adventures, filled with busy days covering lots of distance. He visited surveyors currently working in the field (both for the government and in private practice), retired surveyors, and friends and places from previous years. These trips were an annual highlight for him – he thrived on the opportunity to spend some time away from the city in the more remote parts of the province.

1958

In 1958 British Columbia celebrated one hundred years as a British colony and Canadian province. In his report for the year Deputy Minister Bassett wrote:

> It is fitting the year 1958, which marked the close of the first century of progress in British Columbia, should also witness new summits of achievement by the British Columbia Lands Service on behalf of the people of this province. The volume of activity recorded by the Lands, Surveys and Mapping, and Water Rights Branches of the Service reflect a steadily rising public interest in the land and water resources of British Columbia....
>
> Notable events during 1958 include completion of a two year study of the engineering and economic aspects of the hydroelectric power potential of the Columbia River in Canada. This project ... was undertaken for the purpose of investigating the development of the Columbia River in the best interests of the province. The past year also saw the culmination of a ten year study by the federal-provincial Fraser River Board ... in the form of a preliminary report on flood control and hydroelectric power in the Fraser River basin.

This year, surveyors had the chance to use the tellurometer for field work. Art Barber and Duff Wight attended a one-week course arranged by the federal government in Ottawa on the use of the instrument. In his report Andrews wrote:

> One favourable influence on field operations of the Topographic Division was the full season's use of the new tellurometer equipment for electronic measurement of distances, ranging from about one-half mile to twenty miles [1–30 km].... A full set of readings may be completed and checked for a particular line in a few minutes by a trained crew, which is then ready to move to a new line.... An unprecedented facility for control surveys over extensive tracts of country has suddenly become available.
> It could well have revolutionary effects in methods, speed, accuracy and economy of control surveys.

Both Wight and Barber used the tellurometer for part of their surveying during the summer. One of their projects was a BC Power Commission transmission line.

An increasing demand for large-scale detailed topographic mapping has been experienced and there is every indication that this trend will gather even greater momentum. Horizontal scales vary from 200 feet per inch (1:2400) to 1000 feet per inch (1:12000) with contour intervals from 5 to 20 feet [1.5–6 m]. Such mapping is required for pondage in hydroelectric projects, drainage, irrigation, sewage design, road construction and urban planning.... Without first order plotting instruments or alternatively an increase in trained staff and accommodation, it will be impossible to cope with increasing urgent demands of this kind.

Telurometer in the field.

As the technology related to aerial photography continued to improve, people wanted maps that would provide more detailed information. Andrews realized that his organization needed to continue purchasing equipment that would enable the government to facilitate the ongoing economic development of the province.

He also reiterated a concern from the previous year:

The basic air photo cover program ... was inoperative due to lack of authorized funds to adapt the available government-owned Anson aircraft CF-BCA for air camera installation and to man and operate it. For this reason the northern part of the province between 57 and 60 degrees north latitude still awaits this first prerequisite for effective mapping and development planning. An expenditure of less than $25,000 per year would cover allocation of one photo aircraft to complete the northern British Columbia program in about eight years.

The Surveys and Mapping Branch completed one important survey during 1958. Art Swannell reported:

This year's work on British Columbia's northern boundary ... was not a restoration and retracement survey as it was for the preceding two years, but was primarily an extension from the Tatshenshini River to just west of the Alsek River.... During the summer we had established seventeen monuments and witnessed two other positions along 25½ miles [41 km] of new boundary.

The Boundary Commission reported its decision in August 1957 "to extend the boundary beyond the most westerly monument (166) set by Wallace in 1907, to a point west of the Alsek River. Though this would not complete the boundary, it would prolong it across the last major waterway and bring it to the edge of the enormous icefields that extend to the boundary of Alaska."

Swannell described problems he encountered in the rugged western section:

> Beyond Monument 180, precipitous slopes made it impossible to project the line and very difficult, if not impossible, to reach any point along the line itself. Later, on advice from the Commission, this area was bypassed and points set on the boundary beyond it which depended on triangulation alone for position. At Monuments 181 and 182 the respective sites were witnessed by triangulation stations, the remainder being set in position from stations nearby.

Andrews commented on the significance of completing this boundary survey:

> The new segment extended the boundary westward 25½ miles to a terminal monument No. 187 just beyond the Alesk River. West of this point, to the Alaska Boundary, the 60th parallel is occupied by an enormous icefield, offering no fixed ground for marking the line. Mr Swannell, in establishing Monument 187, may claim then to have set the westernmost survey mark in British Columbia, a distinction which he may enjoy without challenge until such a time as the last remnants of the ice age in the cordillera may have disappeared. It seems fitting that in British Columbia's Centennial Year, 1958, this final operations to mark her boundaries was completed.

There was no summer trip in BC this year for Andrews. Instead, he went on a temporary mission to Southeast Asia at the request of the Canadian government. In a confidential "Memorandum from Secretary of State for External Affairs to Cabinet" on February 4, 1959, Sidney Smith described the project:

> The United Nations Economic Commission for Asia and the Far East (ECAFE) initiated in 1951 a series of field investigations and studies of the Mekong River. The interest of ECAFE was based on the potential benefits which the successful harnessing of the river might be expected to yield to the riparian states in terms of flood control, irrigation, hydroelectric power and improved navigation. For the purposes of the studies undertaken by ECAFE the river includes a drainage area within Laos, Thailand, Cambodia and Vietnam totalling some 235,000 square miles or roughly the area covered by the province of Saskatchewan. This area is commonly referred to as the Lower Mekong River Basin.

At the Geneva Conference in 1954, Cambodia, Laos and Vietnam gained their independence from France. Several countries, particularly the United States, feared that poor economic conditions in Southeast Asia would facilitate the spread of Communism into the region, and they did not want Southeast Asia to become a sphere of influence for Communist China.

In 1957 the United Nations Technical Assistance Administration, at the formal request of the riparian states, appointed Lt. Gen. Raymond A. Wheeler to head a mission with the object of studying and investigating on the spot a number of projects that had been formulated by ECAFE for the development of the Lower Mekong River Basin. The Wheeler Mission submitted their report in January 1958. The report concluded that before any particular projects (such as the construction of dams) could effectively be initiated, further investigations and the collection of basic technical data would be required.

This was the United Nations' first direct involvement in international river-basin planning. Hugh Keenleyside, the director-general of the Technical Assistance Administration, was a Canadian who had lived in British Columbia for several years.

In April 1958 the Executive Secretary of ECAFE visited Ottawa. On that occasion he expressed the hope that, since aerial surveying and mapping appeared to be an important element of the program recommended by the Wheeler Mission, Canada would consider participating in this phase of the program. The Canadian authorities subsequently selected Lt. Col. G.S. Andrews, the surveyor general of the Department of Lands and Forests of the province of British Columbia, to make a detailed study of the problems and estimated costs involved in the proposed aerial survey and mapping of the Lower Mekong River Basin.

As BC's surveyor general Andrews had one of the prominent surveying positions in Canada. He had been a leader in aerial photography for many years, was a member of Fraser River board, and had completed an international surveying mission in 1945.

Andrews spent June through August of 1958 on the Mekong River project. Initially he flew to Ottawa where he met federal officials regarding the mission. He then went to visit the United Nations in New York and travelled to Washington DC to meet American officials involved in the project. From there he flew to Paris and on to Bangkok, Thailand. His guide and host was P.T. Tan, an engineer and ECAFE's representative in Bangkok. Andrews described him in a letter to Jean. "Mr Tan has turned out to be a marvel – he knows all the answers and is a splendid travelling companion. Speaks English, German, Mandarin and Cantonese. His French is not quite as good as mine. He takes me to the best Chinese restaurants – and of course being Chinese – we get the best service – as well as food. He is teaching me to use chop sticks."

Andrews had four main objectives for his mission: First, to gather information regarding existing surveys, maps and aerial photographs and assess their value. Then to make recommendations regarding what further surveys would be necessary, the methods to be used, and the time of year the work should be done. Third, to examine the conditions in the Mekong River area such as access, transportation facilities, communication, and the availability of local supplies and labour. And fourth, to gather information on the geography of the area and the weather, and their effect on surveying.

Andrews travelled up and down the Mekong River visiting all four countries. In a July 16 memorandum he wrote: "A fundamental requirement for the topographic and hydrographic surveys recommended by the Wheeler Mission for the subject river planning is an integrated skeleton structure of vertical and horizontal survey control along the river from sea level to the upper reaches at Burma-Laos border (1600 miles [2600 km])." He noted that there was no common or integrated surveying among the countries. Thailand, the country which had the longest connection with the Mekong, also had the best and most current surveying datum. The French Indochina surveying done in the other three countries was in generally poor condition and did not have the level of accuracy needed for large-scale projects like those envisaged for the Mekong. Thai and Indochina levels were separate and based on uncorrelated tide stations at different locations. But Andrews believed that portions of the French surveying were still usable.

During his travels he had the opportunity to do some sightseeing. A visit to Angkor Wat and a Thai dancing ceremony were highlights. After spending time in the Mekong River Basin he travelled to Hong Kong and Japan and then returned to Victoria in late August where he began writing his report, which he submitted to the federal government.

Andrews described the scope of the project in his preamble:

> The lower Mekong River offers multipurpose benefits of great magnitude to four countries of Southeast Asia which share its regime – Laos, Thailand, Cambodia and Vietnam. Realisation of these benefits is contingent upon harnessing this great river by a series of dams for which several promising and suitably distributed sites have been located. Speculative studies of these dam sites, and the impact of their potential benefits on the economy of the region, have been made by competent investigators, working mainly under the auspices of the United Nations Economic Commission for Asia and the Far East (ECAFE). So encouraging are these possibilities that an International Committee for Coordination of investigations has been set up by the four countries named, to peruse further development on a cooperative basis, under valuable sponsorship of the ECAFE.
>
> The point has been reached, however, where further investigations, planning and design are held up for lack of reliable and detailed data in

The Mekong River.

the form of hydraulic records, precise ground elevations, and topographic maps. It is the purpose of this report to give an appraisal of the surveys and mapping requirements for the next more specific and concrete stage of planning in the development of the Mekong River.

He noted that six potential dam sites had been identified and that they would be used for hydroelectric power, flood control, irrigation and improved navigation.

His report provided specific recommendations regarding the surveying that was needed and how it should proceed. He recommended that the main stem surveying control should be connected to the Thailand surveying datum. Because the Mekong River basin was so large, Andrews emphasized that a "precise levelling is a prerequisite for all the basic engineering data needed." The benchmarks for the levels should be permanently marked and be identifiable from the air photographs. They should also be tied into the horizontal survey as much as possible. Aerial photography was necessary to determine the nature and extent of flooding which would occur behind each dam and to estimate effective water storage capacity. Before the photography could be done it was essential to establish visible control points through the area. This was particularly true for Laos, Cambodia and Vietnam. Andrews recommended that the traverses along the river be done in the most favourable locations, which might entail crossing the river several times. He believed that the dam site areas needed to be surveyed

A market in Nong Khai, a city on the Mekong River in northern Thailand.

more thoroughly while along the placid stretches of the Mekong stations could be further apart. Andrews cautioned that "accuracy specifications for both vertical and horizontal control should be realistic and practicable". It was important to set an order of surveying precision that was achievable throughout the project, not necessarily the highest order used in this type of work.

Andrews agreed with the recommendations of the Wheeler report except for one aspect. Because the Mekong was such a large river and the lower basin covered 2600 kilometres, he felt that it would be more realistic to concentrate activities on the main river instead of including its tributaries:

> The program recommended in this report is directed chiefly to work on the main stem of the Mekong River, which is the core of the whole development. It would seem wise, since the execution of the program depends almost wholly upon aid from sources outside the four riparian countries concerned, to concentrate on the essentials first, especially in view of present uncertainties in international affairs in the Far East as well as in other parts of the world. There is little doubt that the successful completion of the proposed program, say by end 1960 or 1961, if tempered by a reasonably stable international outlook at that time, will constitute a strong case for further activities.

He concluded by saying he believed that the surveying and mapping of the Mekong River Basin was "a key step in the realization of the ultimate physical modifications to the river which will unlock its great potential service to the region".

In his memorandum to the federal cabinet, Sidney Smith wrote:

The Mekong project appears to be a good example of a co-operative project that will yield benefits to more than one of the countries of the region.... The countries directly concerned with the Mekong project ... are all members of the Colombo Plan. Three of these countries are also countries with which Canada has formed particularly close contacts through our service on the International Truce Commissions in Indochina.

Since Canada was also a member of the Colombo Plan, Smith recommended:

a) The Canadian Government agree, in principle, to undertake the first priority phase of the surveying and mapping of the Mekong River at a cost not to exceed $1.3 million.

b) The sums required for this purpose be met from Colombo Plan appropriations for 1959–60 and 1960–61.

c) Canadian participation in the project be subject to the submission of a joint formal request for Canadian assistance by Laos, Thailand, Cambodia and Vietnam.

d) The recipient Governments be informed that Canadian assistance in the first priority phase of the project implies no commitment on the part of the Canadian government to participate in any subsequent or ancillary phase of the project.

e) The Executive Secretary of ECAFE be informed of the cabinet's decision in the foregoing terms.

Cabinet accepted Smith's recommendations and a Toronto company received the contract as Management Engineer for the Mekong project. Two BC surveying companies did part of the work. But the Vietnam War ended any proposed development of the Mekong River, and it was not until 1995 that the four countries signed an "Agreement on the Cooperation for the Sustainable Development of the Mekong River Basin". China and Myanmar (Burma), the two countries on the upper Mekong, later became affiliated with this agreement.

As a follow-up to his overseas experience, Andrews provided training opportunities from two to six months for four overseas students during the next five years. Two of the participants came from Thailand. In his 1964 report Andrews wrote:

A pleasant and interesting variation in the normal routine of this branch has been provided on several occasions by the attachment, for observational training, of a student or fellow from far away, under the auspices of the United Nations or of the Canadian External Aid office. The most recent of these, sponsored by the latter, was Miss Chaleuysri Siricharoen, B. Comm., from Thailand, who was interested in cartography,

photogrammetry and statistical geography. Miss Siricharoen spent July
and August of 1963 and May of 1964 with this branch, the intervening
period having been spent in academic studies at the University of British
Columbia.... It is a pleasure to report that all of these assignees made a
very favourable impression among us by their intelligence, conscientious
application to the study program, as well as personal charm.

Andrews kept in touch with P.T. Tan and his family for many years.

1959

Although 1959 was a largely uneventful year, Andrews described the role of
the Surveys and Mapping Branch in development projects throughout the
province that year.

Headlines of British Columbia newspapers did not lack for sensational
and historic material during 1959, and although surveys and mapping, as
such, did not feature among these, it is a fact that the underlying but mute
basis for much of the dynamic news of today was indeed the result of
fundamental survey services of various kinds.

In connection with the visit of Her Majesty Queen Elizabeth II, the
Geographic Division prepared and supplied special maps upon which
the important details of the Royal itinerary were assessed and finalized.
Tactical details of the ceremonies at the Provincial Parliament Buildings
in Victoria were worked out effectively on the basis of large-scale detailed
vertical air photographs especially taken for the purpose by the Air Division. Similarly, from air negatives held by the same division, special photo
enlargements of the Pennask Lake locality were promptly produced, under
strict security, to assist those responsible for the weekend of quiet rest and
recreation for Her Majesty and party in that beautiful sylvan retreat.

In connection with sensational hydroelectric power proposals now
at issue in British Columbia, and so prominent in the news, it should be
remembered that no firm estimate of power at any site or for any system
of sites is possible without accurate detailed topographic surveys of the
terrain at and behind each proposed dam, for determination of water storage and pondage potential, and for installation design. Accurate elevation
data are essential for estimating the potential "head" and "draw down" at
each dam. The Topographic Division, using air photos obtained by the Air
Division, has done the field control surveys and the compilation of detail
topographic maps for much of this type of work on the Fraser, Homathko,
Peace, Stikine, Liard, and other British Columbia river systems, at the request of the Comptroller of Water Rights, the Fraser River Board, and the
British Columbia Power Commission.

The stock-taking, management and fire protection of our forest tree-farm licenses, public working circles [and] access roads – always front-page news in British Columbia – depend on maps and air photos for the effective resolution of decisions concerning them.

Provincial revenues have been handsomely fattened by the disposal from time to time at public auction of petroleum and natural gas rights on Crown lands, especially in the vast northeast triangle of British Columbia. The effective smooth-working administration of these tenures is based on a unique grid system of permits and leases, based on geographic co-ordinates which was conceived and worked out in detail by the Surveys and Mapping Branch....

Returning now to the "fish vs power" feature of current British Columbia news, in addition to the survey services supplied to the hydropower aspects already mentioned, the experimental, operational and consultative services have been sought by, and freely given to, the various fisheries agencies, particularly in the application of air photography for spawning studies of herring on the coast and salmon in the up-country streams.

And so, when the spotlight of prominence illuminates those connected with the big business of British Columbia – forests, hydropower, flood control, fisheries, transportation, etc. – in the press, radio and television, it should be remembered that among the fact finders behind the scenes a vital and indispensable part has been played by those concerned with surveys and mapping, without prejudice, fear or favour.

The Surveys and Mapping Branch had one major development in technology during the year. "The installation of a Swiss-made Wild A-7 autograph plotter in the photogrammetric section of the Topographic Division was a major step in the modernization of BC government surveys and mapping potential. This is a stereo plotter of first order precision of worldwide renown." The tellurometer was used for surveying around Clinton and in Wells Gray Provincial Park. Ken Bridge used two receivers for his work this year, enabling him to measure more distances with the tellurometer.

Art Swannell spent one more field season on the BC-Yukon Boundary Survey:

At a meeting of the [Boundary] Commission it was decided that an investigation and retracement of Mr Wallace's 1907–08 survey should be made. The investigation was to start at Monument 166 and extend eastward until the possibility of discrepancies was removed. The decision was due to the findings of Swannell given in his 1958 report and which were substantiated by a tie of the Army Survey Establishment triangulation network to Monument 148 on the Haines Road. The overall distance between Monuments 166 and 141 had shown a substantial discrepancy whereas that between Monuments 141 and 120 was satisfactory.

One of Swannell's crew members at Monument 141.

Swannell had packhorses for transportation and a tellurometer for mea-
suring distances. Regrettably, poor weather prevented the completion of the
work. In his conclusion Swannell stated:

It is unfortunate that the entire retracement was not completed. The limit-
ing factor was the allotted funds which would, with ordinary achievement
based on three previous season's costs and daily output figures, have been
adequate to complete the project. But serious and lengthy delays due
entirely to adverse weather, retarded the work beyond all premeditation....
The portion of the boundary retraced this season ... has been surveyed
to the required standards of accuracy. In all, 24½ miles [39.5 km] were
retraced. This leaves 11¾ miles [19 km] which have not been retraced.

In 1960 the Boundary Commission decided that "further expenditure
was not justified for so little work in so remote an area", completing the
survey of the 60th parallel in BC. It reported:

On August 15, 1963, a ceremony was held at an aluminum monument which had been erected to mark the completion of the boundary
between the provinces of Manitoba and Saskatchewan and coincidentally,
the final demarcation of the 60th parallel of latitude separating British
Columbia, Alberta, Saskatchewan and Manitoba from the Yukon and
Northwest Territories. Spanning more than 1500 miles [2400 km] from
the Panhandle of Alaska to the shore of Hudson Bay, it is thought to be
the longest visibly marked parallel of latitude in the world. Federal and
Provincial government officials attending the ceremony witnessed the de-
position of a time capsule containing various objects and letters addressed
to the government officials in the year 2063....

The boundary between British Columbia and the Yukon and North-west Territories, which is the concern of this report, covers nearly 661 miles [1064 km] of the 60th parallel, although only 622.2 miles [1001.3 km] have been surveyed and monumented. Most of the remaining 38.7 miles [62.3 km] is covered by glaciers.

In 1966 the British Columbia-Yukon-Northwest Territories Boundary Commission submitted their final report (Andrews was still one of the commissioners), and in 1967 the BC government passed An Act of Consent Respecting the Adoption of the British Columbia-Yukon-Northwest Territories Boundary.

1960

The development of water projects was one of the main areas of involvement for the Surveys and Mapping Branch in 1960. Deputy Minister of Lands Ed Bassett wrote:

> During the year the report of the Peace River Power Development Company concerning plans of this company for the harnessing and transmission of power from the Peace River was submitted to the comptroller. As in the previous year the Water Rights Branch also participated on a technical and advisory level in negotiations between Canada and the United States for the development of the Columbia River....At the end of 1960, the Canadian and United States teams were very close to agreement on the draft of a treaty.

Surveys were also made for potential water projects along the Stuart River waterway in central BC, the Liard River, and in the Bowron Lakes area.

The branch worked on several projects for industries not concerned with water. It carried out small mapping projects related to the field surveys done for the British Columbia-Yukon-Northwest Territories boundaries. Andrews reported:

> The petroleum and natural gas industry in northeastern British Columbia has been assisted during the year by a number of survey ties between the Alaska Highway right-of-way monuments and the broad triangulation network....The numerous monuments of the Alaska Highway survey can now be coordinated on the North American Geodetic Datum of 1927 and thereby will be of much greater value to the oil industry for many well-site locations having access from the highway.

The largest technological advancement in 1960 came when the provincial government installed an IBM 650 computer that provided "unprecedented facilities for survey computations of many kinds". The Fraser River Board supplied a second plotter for the Survey Division to use for the

Charlie Wise (left) and Jack Aye.

mapping work they were doing for the board. Andrews wrote about the impact that technological upgrades had on his department:

The helicopter, the float plane, better roads and vehicles for speedy deployment to, in, and out of the work areas; field radio communication for better tactical coordination and weather intelligence; the air photos and interim maps for planning and appraisal of the various field projects; the tellurometer for speeding up and improving accuracy of distance measurements; and the precise air-photo plotting equipment which yields standard accuracy and detail of mapping, on a greatly reduced density of fixed ground points – all have combined to reduce the effort required on field control surveys in proportion to the compilation and drafting of plans.

He also wrote about the possibility of integrated surveys, an issue that would become important for him later in the 1960s:

Field surveys – for controlling the rigorous standard topographic mapping (with contours); for serving the needs of the petroleum and natural gas exploration; for the demarcation of provincial boundaries on the east and on the north; for highway rights-of-way surveys across the south, up the middle, and on through the north – using triangulation, tellurometer, and precise traverse techniques, have resulted in an imposing structure of survey control well coordinated with the North American geodetic datum. This provides the framework for integrating all surveys of the future.

Many people who knew Gerry Andrews remember the multitude of friendships that he kept over many years with a wide range of individuals, from prominent people to packers. They also comment that he continually remained interested in BC history and liked to revisit places where he worked. One of Andrews' life-long friends was Jack Aye, the packer on his 1930 Flathead crew. In early July 1960 he wrote to Aye:

> Can you take a couple of days off, say between Monday the 18th of July and Wednesday the 21st of July? Why? This is why: I have to be in the Cranbrook vicinity about the 16th or 17th, with a four-wheel drive station wagon, and if I could persuade you to join me, would take a couple of extra days to drive down through the Flathead on the forestry road via Corbin – as far as the lower end. I could pick you up at your place complete with bed roll and fishing rod – say on Monday morning the 18th – and deliver you back to your beloved women-folk, say on Wednesday evening the 20th. I would have to fly back to Victoria on Thursday the next day from Cranbrook. Don't you think we owe ourselves this little trip after 30 years? If you can, drop me a card or a note right away, as I will be leaving here on the first part of my trip early on Wednesday the 13th of July.

Aye immediately sent Andrews a telegram accepting the proposal, and the two men spent three happy days revisiting the Flathead, including a visit with trapper Charlie Wise, who still lived in his cabin there.

1960s

During the 1960s the Surveys and Mapping Branch remained busy, but the hectic pace of the post-war years began to slow down. There was still a high demand for Crown land to be surveyed, but there were fewer large surveying projects. Increasing use of technology meant more automation in the office and fewer surveyors in the field. In his 1961 report Andrews wrote:

> In several forms, automation has already been an established feature of field operations, especially in control surveys for topographic mapping – namely, air transport, radio communication, weather prognostication, air-photo bridging, as well as in the fundamental tasks of measuring distances and angles. Due to these, it has been possible to cope with increasing demands for such surveys with little, if any, increase in field staff. In fact, it may be remarked with regret that we are no longer able to offer summer employment to students during their vacations, to any significant degree.

Remembering his experiences as a young man, Andrews wrote: "This effect of automation also denies these young men the physical and moral training, unique to survey work, especially in the wilderness, which would be so beneficial throughout their lives, no matter what calling they may follow later." In his 1962 report he noted that there were 174 employees in the Survey and Mapping Division compared to 192 a decade earlier.

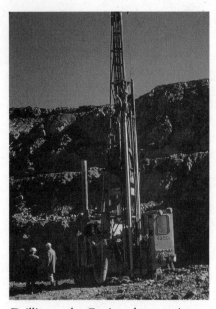
Drilling at the Cassiar asbestos mine.

The use of computers for office work increased in the 1960s. Andrews' 1961 report noted "the adaption of various survey calculations to speedy and faultless solutions on the IBM 650 electronic computer recently installed by the provincial government. By this means lengthy and involved computation operation for checking adjustments, conversions, etc. are now a matter of routine." It also mentioned: "A prominent member of our staff having high mathematical and survey qualifications has developed and perfected a number of original programs for the IBM 650 computer." And more specifically, "by arrangement with the Attorney-General's department, on a trial basis, all subdivision plans submitted to the Land Registrar, Victoria, have been checked with the 'Surmap 03' program with such good results that consideration is now being given to extending this service to the other Land Registry Offices." In 1962 the service extended to the Prince Rupert, Kamloops and Nelson districts and "a satisfactory work flow routine was established, whereby plans received in Victoria by Monday's mail were checked and in return mail by the following Friday."

The branch also upgraded equipment in the field:

One of the two Anson aircraft (CF-EZN) acquired in 1949 was replaced successfully by a more modern all-metal Beechcraft (D18S Expeditor, CF-BCE). A second identical replacement unit (CF-BCD) was acquired at the same time, anticipating retirement of the second Anson (CF-EZI) during 1962 or soon after. The Ansons have given outstanding service to the province during the sixteen-year post World War II period to date, having photographed a gross of nearly 500,000 square miles, (including revision projects) recorded on some 300,000 high-quality air photo negatives.

In 1963 the venerable CF-EZI was retired after 7,152 hours of flying during which it travelled over 1,600,000 kilometres.

Transportation difficulties hampered the Surveys and Mapping Branch in 1961. "Due to this changeover from the Anson CF-EZN to the Expeditor aircraft CF-BCE not being effective till well in the season the completed program was, for the most part, restricted to one operational Anson." One

Pinchi Mine.

of the chartered helicopters was not operational for a while. And,

> early in May, while the Department's DeHaviland Beaver was based at
> Port Hardy,... a fish-boat collided with the plane, which was moored at
> the Department of Transport float, and tore off a wing. The Engineering
> Section of the Forest Service cooperated immediately by dispatching its
> self-propelled barge from Vancouver to transport the crippled plane to
> Victoria, where the necessary repairs were to be made.... An insurance
> claim is presently in the course of being settled.

The major surveying in the field was related to water: "Hydroelectric powers on northern streams were continued, with large-scale mapping projects being completed for the Stikine and Liard rivers.... [And] sixteen years of study and negotiation culminated in the signing of the Columbia River Treaty by Canada and the United States on January 17, 1961."

During the 1960s restoration surveys became increasingly important and were an ongoing concern. A 1947 regulation required all new survey markers to be permanent monuments. By the 1960s many of the original survey markers, which often used wood, could no longer be located. "The purpose, as the name implies, is to renew important survey monuments which, by the ravages of time, have disappeared and thus impose an unfair burden on citizens having interest in nearby land when resurvey becomes necessary in connection with safeguard or transfer of title, subdivision, trespass, etc."

Integration of surveys continued to gain importance during the 1960s, particularly with the growing number of surveys in the province. Andrews explained the importance of integration and the process:

The Peace River dam site in 1962.

Reference has been made to the integration of all surveys on the controlled North American Datum (1927) by coordination of all monuments or markers governing the legal boundaries of property, engineering projects, bench marks and indeed man-made structures of a substantial and permanent nature with the survey control network....

Integration of surveys has received increasingly widespread support in recent years. It will be the theme of the forthcoming annual convention of the nation-wide Canadian Institute of Surveying in Ottawa, February 1963.... It is only in recent years that the necessary overall control surveys have been extended sufficiently to provide a widespread structure for survey integration. Also, it is only recently that certain revolutionary technical equipment has been available to make practicable for the first time in history the rapid and economic breakdown of the primary control structure into a density of stations necessary for effective universal survey integration.

These technical developments come mainly from electronics as applied to distance measurements and computation, from precise aerial photogrammetry, from air-borne field transport, and radio communication. Even with these new technological aids, the task of catching up with the integration of the enormous aggregation of cadastral surveys from the past is colossal.

The integration of surveys would be done by a series of monuments visible to each other and all connected to the North American Datum. These monuments would be surveyed to a very high level of precision. All surveys would be tied into at least two of these permanent monuments.

By 1962 the helicopter had become the main method of transportation to the triangulation stations on remote mountain peaks.

The density of control monuments will vary according to the development of an area. In cities they must be spaced about 1,000 feet [300 m] apart, that is, 25 per square mile [10 per km²]. The suburban density may thin out to four or five and rural to about one per square mile. However, if an average of say only three monuments per square mile were struck for the whole of British Columbia, it would imply over 1,000,000 monumented control points, over 30 times the number we now have....

The integration of surveys has been a feature of British Columbia survey policy for several decades, but due to the enormity of the task, catching up with more or less a century's lead in primary lot surveys ... and subdivision and right of way surveys ... only sporadic ties could be made between the new expanding mapping control system and nearby district lot corners, as opportune. In spite of these limitations, a very creditable array of coordinated points has been established. Some 33,000 are now documented in our control records.

[During 1962,] with the gratifying approval of the Honourable Minister of the Department, a modest beginning on a specific program for integrated surveys in British Columbia was undertaken during the past field season in the Lower Fraser Valley. Due to difficulties with weather, smog, and rather expensive helicopter access to key mountain geodetic stations, only 23 new stations were established instead of some 40 originally planned.

The plan was to bring integrated surveys into one area at a time. Once there were sufficient monuments, all future surveys in that area would be integrated into the network. Andrews felt that the results from the first season

Andrews' 1963 Trip

Karl Hanawald's abandoned trading post.

During his travels through northern BC in 1963, Andrews visited the First Nations village at Bear Lake in the Omineca Mountains, deserted since the early 1950s. There, he saw the abandoned trading post that had been operated by well-known trader Karl Hanawald from the early 1930s to around 1950. Andrews then flew by helicopter to nearby Mount Coccola to inspect the surveying being done there. From there he flew to the Big Kettle fumarole on the Omineca River where he located a blaze left by Frank Swannell's survey crew 50 years before. He also took a photograph of the fumarole from the same location that Swannell did in 1913.

The crew on Mount Coccola uses a tellurometer (right) to measure distances.

Andrews at Swannell's blaze.

The Big Kettle fumarole.

provided valuable experience and that his department was now ready to establish a concentration of integrated survey stations in selected areas.

In 1963 the program in the lower Fraser River valley continued, focusing on Surrey and New Westminster as the first areas for survey integration. Progress was slowed because the Surveys and Mapping Branch was unable to obtain in time one of the new tellurometers that could measure the shorter distances involved in urban areas. In a winter works program the Surrey engineering department constructed 200 control monuments to facilitate the 1964 surveying.

At the end of 1963 the Fraser River Board made its final report and in early 1964 it disbanded. Although pleased with the work accomplished, Andrews also had a feeling of relief, for he had spent a large amount of time over several years as a member of the board.

> There is considerable satisfaction in the debut of the final report of the Fraser River Board coincident, for practical purposes, with the end of the year. It marks the end of more than eight years' service as a Board member for the writer, and parallel service as alternate member for Mr A.H. Ralfs, BCLS, DLS, Assistant Director of this Branch. Whereas participation in the comprehensive analysis of this great river's characteristics and potentialities, and the synthesis of a plan for its regulation to effect flood control with production of hydro-electric power, has been a unique privilege, it has nevertheless imposed a very real diversion of attention and energy from responsibilities intrinsic to the Surveys and Mapping Branch....
>
> In addition to provision of services as member and alternate, this branch has performed a substantial aggregate of special assignments in detail topographic mapping, including special air photo cover, of the numerous reservoirs and dam sites required for the board's studies. The precise determination of a critical elevation on the Parsnip-McGregor divide at Arctic Lake offered a major feature in the Board's planning for the diversion of the McGregor River waters into the Peace River drainage basin.

By the time the Board made its report, the construction of dams on the Columbia River had already started, and there was increasing interest in dams in northern BC. In 1964 "field control was also conducted in the Peace River power project reservoir area." This continued the following year when "five of our personnel were engaged on a three month control survey on the Peace River power project." Surveys for potential dam sites on the Liard, Kechika, Fort Nelson and Stikine rivers were also done in the mid 1960s.

In 1964 a legislative bill was passed for survey integration:

> The Bill, in the form of an amendment to the *Official Surveys Act*, provides for the Proclamation by Order in Council of integrated survey areas within duly defined boundaries, when, upon the recommendation of the

W.A.C. Bennett dam site in 1964.

surveyor general, sufficient survey control monuments have been suitably
established and co-ordinated. Thereafter, all new legal surveys under the
Land Registry Act within the said area must be co-ordinated to the common control.

Andrews considered the establishment of Integrated Survey Areas one of his
key accomplishments as surveyor general.

The selection of areas for survey integration will depend mainly on the
order in which bona fide willingness to participate is shown by each local
community; that is first come, first served. However, for the major control
operations, concentrated groups of individual local authorities would be
handled as a group unit, with sufficient flexibility for an individual area to
complete its part of the program as and when practical, from budgetary
and other considerations. Obviously the spread of survey integration to all
communities of the Province will take some time to achieve.

Andrews remarked that awareness and interest in survey integration was
spreading throughout the province, but he added a note of caution: "Experience with Surrey ... indicates that the setting up of adequate major control
for integration in the Greater Vancouver area is such a large job that assistance from the Federal Department of Mines and Technical Surveys should
be sought."

The Surveys and Mapping Branch acquired new equipment to further increase the department's efficiency in 1964. It purchased a new air camera with a special lens, along with two tellurometers and a high-precision theodolite. New computer programs were developed to accompany the IBM 1620 computer purchased the previous year. In 1965 the purchase of a Model 6a geodimeter, developed in Sweden, solved the difficulty of measurements that were under a mile (1.6 km). "The geodimeter is thus ideal for close work in urban and congested areas."

In his 1964 report the Deputy Minister of Lands noted a development involving surveying that was increasing around the province:

> The province's industrial prosperity is spreading in an ever-widening radius away from the traditional core in the major cities. Expansion of the mining and forest industries has stimulated the alienation of blocks of Crown lands as the basis for the private planning and development of entirely new communities. Examples of such new communities based on mineral resources are at Hendrix Lake, Topley Landing, Fraser Lake, and Gowling Island, while Rumble Beach and Gold River have been founded as the result of the growth of the forest industry.

In his 1965 report Andrews wrote:

> Contrary to the hope expressed in my previous report, the year has ended without the first "integrated survey area" being proclaimed under provisions of the *Official Surveys Act* as amended in 1964. A number of areas are, however, almost ready for this historic step, their major control networks having been installed, surveyed, adjusted, checked and double checked to the required specifications.

In 1966 Surrey became Integrated Survey Area No. 1, followed by Dawson Creek.

The Surveys and Mapping Branch returned to the oil and gas fields of northeastern BC in the same year:

> The Topographic Division embraced a varied and somewhat expanded field program which included a return to the muskeg country northeast of the Alaska Highway after a lapse of some 11 years since the original triangulation program there, this time to initiate improved vertical control with spirit levels ... and to upgrade and intensify the horizontal control, with the advantage of tellurometer for trilateration and traverse, and the new first-order control net established by the Geodetic Survey of Canada to strengthen the structure north-east of Fort Nelson. This project is tailored to the requirements of accelerated drilling for petroleum and natural gas in that area, and confirms the wisdom of our original policy in 1953, 1954 and 1955 to cover the whole area first with an extensive but sparse net of control, anticipating further work when and where favourable areas for potential exploitation became known. The experience gained in this year's

project will benefit a much larger scale of operation of this kind anticipated in the 1967 season.

This survey involved more work on the ground than the original one:
Over the soft ground and muskeg sections, six-foot aluminum poles with
adjustable sliding brackets were used as turning points and for instrument
set-ups, which proved invaluable and were designed and constructed in
the Air Division workshop before the crew left for the field, utilizing our
previous season's experience. This portion of British Columbia, with its
gumbo mud and muskegs, is a very unpleasant part of the province for
surveyors working there.

In the summer of 1967 the branch sent a large survey crew to work in
northeastern BC. Andrews wrote:

The Topographic Division had a record year in respect to the magnitude
of its field operations and the budget to finance them. This situation
arose largely from a crash program to improve the quality and density of
horizontal and vertical control in northeast British Columbia in response
to representations from the Canadian Petroleum Association, and was approved on our recommendation that it would be more economical to get
most of it done in one season by an enlarged operation than to do it in
piecemeal for a number of years.

Unfortunately this large survey ended two weeks early due to a government
austerity blitz and failed to complete its objectives.

The summer of 1967 was the busiest one in Andrews' last years as surveyor general. In addition to the large project in the oil and gas fields he
had a sizeable survey crew working on mapping along the Stewart-Cassiar
Road. Through the years sections of the Western (Cassiar) Highway were
being constructed, although the route was not open to the public until late
1972. Work related to dams was being done both on the Arrow Lakes and in
the Peace River area. For the Columbia River dams, "we were committed
on a cooperative effort with the Geodetic Survey of Canada and designed a
triangulation network between the existing geodetic stations at Trail, Cres-
ton and Revelstoke." A second project "consisted of establishing a network
of stations by tellurometer traverse above the take line of the proposed
flooding of the Arrow Lakes to aid in the preservation of existing cadastral
surveys.... The traverse stations so set will also be a basis for future cadastral
surveys following the flooding." On the Peace River there were surveys for
"three dam sites on the Peace River downstream from the W.A.C. Bennett
Dam".

The integration of surveys project was also active. The federal Surveys
and Mapping Branch provided assistance in establishing stations in both the
Vancouver and Victoria area. Several communities in BC started preliminary work on survey integration.

In his 1967 report, his last as surveyor general, Andrews reflected on the large number of technological changes that had occurred during his time:

> Features of conspicuous but progressive change during this period are technological, covering equipment and methods, happily with a corresponding increase in the quantity and quality of work output.... The technological features mentioned include improved photo aircraft, the cameras and the navigational aids used in them, together with processing facilities. Revolutionary procedures for distance measurement in field surveying became available in the late 1950s in the form of the tellurometer and the geodimeter, as did first-order photogrammetric plotting equipment in the form of the Wild A7 autograph. Complementary with these, the revolution in computations and data processing by electronic systems was propitiously and progressively used, such that almost immediate advantage of the improved field measurements in the form of sophisticated adjustments has made practical reality of what was formerly theoretic and wishful thinking. Improvements in cartography, lithography, and in drawing and reproduction materials, particularly in respect to dimensional and chemical stability, have been significant. It has only been by persistent and aggressive adaption to these new advantages that the static staff potential has been able to cope with the compounding increase in volume, variety and complexity of demand.

At the same time, his reports acknowledged almost every year the importance of the people who worked in his department and the value of their skills.

Two projects in 1967 illustrated the complexity of demand the Surveys and Mapping Branch faced in their requests:

> Among the items requiring considerable time and technical competence was the system developed for the Department of Highways to record highway features on 16 mm film from a moving vehicle.... Another highly technical and specialized project was the work done on the Forest Service Linhof 70-mm cameras for stereo-photography from a helicopter.

Canada's centennial year was 1967, and Andrews was involved in an interesting nationwide project to recognize the importance of surveying in the country's history.

> At 11:15 am (Pacific Daylight Saving Time) on June 21st, the day of the summer solstice, British Columbia participated in simultaneous nation-wide Centennial ceremonies, paying tribute to surveyors for their work in measuring and mapping Canada during its first centenary, 1867 to 1967, and the establishment during that period of the North American Geodetic Survey Datum across the length and breadth of our nation. The ceremony took the form of dedicating the site of British Columbia's Canadian Centennial survey monument in a selected spot in the precinct of

Granduc Mine.

the new Provincial Museum in Victoria.... Simultaneously, similar Centennial survey monuments were dedicated in all the provincial capitals across
the country, in Whitehorse, Yukon Territory and Ottawa.

The monument was a brass-cap fixed survey point that was tied into the
North American Geodetic Datum and could be used by future surveyors. A
commemmorating plaque was also placed there, but was later removed. All
of the centennial survey monuments had similar features. Andrews was the
master of ceremonies for the event. Attorney-General Robert Bonner gave
the coordinates of British Columbia's survey monument to the audience
and joked that he was the only politician who knew exactly where he stood.

Because there was a large amount of field activity this summer, Andrews
took a long trip to visit the crews in the field. When he was in northern BC
he went to Whitehorse to view their centennial monument.

Andrews retired as surveyor general in the fall of 1968, so the annual report for that year was written by his replacement, Bert Ralfs. Andrews did
not go on a summer trip in 1968, but made three short trips to northern
BC to visit some of the projects in the region. These included the Granduc
mine, the Granisle mine, the Bennett Dam construction and the flooding of
the Finlay River. Even in his final year Andrews continued to incorporate
new technology into surveying in BC: the first use of polaroid cameras; a
new radio repeater system that provided more immediate communication

in the field; new computer programs to increase office efficiency; and better equipment to improve mapmaking. Work continued on the Integrated Surveys Area, his major project in his final years.

Gerry Andrews served as surveyor general for 17 years, the longest anyone has held this position in British Columbia. During that time British Columbia had the fastest growing population in the country. It was a time of tremendous economic development that transformed the province. In his role as surveyor general and head of the Surveys and Mapping Branch, Andrews and his staff played an active role in almost all of the major projects in British Columbia during these years.

Retirement

Gerry Andrews started his retirement in the fall of 1968 by taking a lengthy solo trip down the west coast to Mexico, a country he had previously visited a few times. Shortly after returning he found that his surveying expertise was still in demand. The Canadian Advisory Council on Cadastral Surveys, comprising the chief legal survey officers of the federal and provincial governments, requested a study of survey administration across the country. The federal Department of Energy, Mines and Resources funded this request and selected Andrews to undertake the study. Between mid January and mid March 1969, Andrews met with the leading federal and provincial survey authorities in Ottawa and each provincial capital. His report, submitted in 1970, provided an overview of the legal survey system in each of the provinces. Since there was a variety of well-established cadastral survey systems in Canada, Andrews did not recommend any major changes.

While he worked on this project, Andrews received an offer from the United Nations' Cartographic Branch to spend a year working on a photogrammetry project. It was a special opportunity for him, but the timing was not right. He had only just retired and did not want to leave Victoria for a year, especially while he was still writing his report for the Canadian government. He received some more offers for employment outside Canada around the same time, but turned them down too. Later, in 1972–73, he did take a six-month overseas assignment for the Canadian International Development Agency, teaching photogrammetry at the Federal University of Paraiba in Brazil.

Andrews also travelled overseas for pleasure. In the summer of 1969 he went to Europe with Jean and Kris, who had just graduated from university. Two summers later he travelled alone to the USSR, then in 1975 to Greece, Crete, Istanbul and other places in the eastern Mediterranean, and in 1978 to Spain and England. In 1981 he visited China and went to the hospital in

Chungking where his aunt had worked as a nurse from the early 1920s until 1949. He took her photo album along, and to his surprise and delight some of the older nurses recognized themselves in the pictures. Finally, in 1986, he went to Peru, where he visited Machu Picchu and other sites; but the main attraction there was a clear view of Halley's Comet, which he remembered seeing in his childhood.

In retirement Andrews became active in studying the history of British Columbia. He served as president of the BC Historical Society from 1972 to 1974 and wrote several magazine articles for the society's magazine, all having some connection to work he did. The articles gave him an opportunity to reminisce and reflect on his involvement in historical events, to correspond with old friends and colleagues connected to them, and to go back to the areas where they took place. Andrews also remained in contact with the Association of BC Land Surveyors. He wrote a few articles for its magazine, *The Link*, obituaries for several surveyors he knew, and a booklet that listed all of the province's surveyors and noted any special surveying assignments they completed.

Andrews spent considerable time in the early 1980s researching and writing *Metis Outpost*, a book about his two years of teaching in northern BC, published in 1985. He also gave many public presentations, including several at the Royal BC Museum. In 1989 he travelled to Vanderhoof where he spoke for the dedication of the old Sinkut Astro Pier surveying monument at the museum on Canada Day. That fall he went to Kelly Lake to speak at the dedication of a new school in the community and to introduce another speaker, the daughter of his old friend, Jim Young, who had organized the first school there.

Most of Andrews' travels during retirement were in western Canada. In the fall of 1970 he returned to northern BC. He travelled to Big Bar, drove through Pine Pass, visited the Bennett Dam near Hudson's Hope, and spent time in the Peace River area. Like the trips he made as surveyor general, he visited the places where he had worked in previous years and he renewed friendships with many people. The next fall he went to the areas around Lillooet, the Goat River and Valemount. In June 1972 he and Al Phipps (who had worked in the Surveys and Mapping Branch for several years) travelled to northern Vancouver Island, visiting places connected to Andrews' 1934 forestry work. In September he canoed the Bowron Lakes chain with George Newell and his wife, whom he had met through the BC Historical Society. Newell remembers Andrews taking him to visit retired surveyor Harold Garden in his cabin at Barkerville after the trip, and the mutual delight the two men had with their visit.

In November 1972, the Stewart-Cassiar Highway (no. 37) was officially opened to the public, so in the summer of 1973 Andrews returned to Atlin

Andrews in Atlin.

for the first time in several years. In a letter written much later, he explained his connection to the area:

> I flew over Atlin in 1948 taking air photos – then in 1953 and frequently later visited the area in connection with my job as surveyor general and director of mapping. In 1965 a colourful old Atlinite, Harper Reed, died at 87 and left me his cabin there. Ever since they opened the Cassiar Road north from Hazelton in 1973 I have driven up and enjoyed my Atlin cabin – a wonderful retreat. About that time they made me a "lifetime" member of the Atlin Historical Society.

In declining health, Reed had moved to Victoria in 1963 where Andrews visited him often. Reed was a bachelor with no immediate family members. When he died Andrews was surprised and pleased to learn that Reed had left his Atlin property to him.

Almost every summer from 1973 to the early 1990s Andrews spent three or four happy weeks at the cabin in Atlin, enjoying the rustic life, touring around the area and visiting with the people there. Friends often accompanied him to or from Atlin and stayed with him for part of the time. Some would fly to Whitehorse and then drive south to visit him. In the summer of 1973 Al Phipps joined Andrews on the drive north from Victoria. They picked up George Newell and his wife in Hazelton, and then Duff Wight. This was Newell's first trip to northern BC. While he enjoyed the scenery, Andrews and Phipps scanned the mountain tops for survey monuments and remembered the surveyors who had worked in the area. Newell was amazed

Andrews surveying at Atlin.

by the energy Andrews displayed on these trips. One day, when they arrived at Dease Lake in the early afternoon, Andrews decided that they should go to Telegraph Creek for a quick visit, even though it was more than 110 kilometres away on a gravel road.

Jon Magwood remembers one of Andrews' trips to Atlin in the 1970s. Magwood was surveying at Dease Lake when Andrews and Lorne Swannell stopped to visit. Andrews was going to survey his property in Atlin and he asked Magwood for a few pointers, since he had never done a legal survey before. Andrews brought out his red sock and flask to share while they talked. Magwood reminded Andrews that he had been using the book of surveying instructions signed by G.S. Andrews for many years. Andrews pointed out that, even though he had signed the document as surveyor general, he hadn't written it. Later, on his way south, Andrews again visited Magwood and brought out his red sock. He expressed concern that he had run out of surveying pins and had to use angle iron for one of the survey corners. Magwood reassured him that if he marked "angle iron" by the pin used and put his initials beside it no one would question the validity of his survey.

Andrews' daughter, Kris, who lived in Williams Lake, accompanied her father a few times on his Atlin trips. She and her father also took a few short trips through the Chilcotin.

Andrews occasionally took trips through eastern BC and to the Prairies. These trips, usually for two or three weeks, covered about 5000 kilometres in a loop that would allow him to visit several places and people, often members of his family, especially his sister Leila in Gull Lake, Saskatchewan, and his daughter Mary in Banff. His Ford Econoline van served as a camper and he would often stay at municipal campgrounds. He liked to travel with a friend, but in 1974 Al Phipps died and George Newell moved to Prince Rupert, so different people joined him on his Prairie trips.

Andrews also kept in touch with many friends that he had in the Peace River region. Sometimes he would visit there on a return trip from the

Andrews with daughter Kris in the Chilcotin.

Prairies. In the summers of 1981 and 1982 he drove back from Atlin along the Alaska Highway to the Peace River.

In 1981, he also went to the Chilcotin with Kris. And in 1983 he spent all summer travelling through Alberta, along the Alaska Highway, in Yukon, up the Dempster Highway and around Atlin. It was on his return trip at the end of that summer when I first met Gerry Andrews.

In 1984 he wrote to his old forestry friend Hugh Hodgins:

> My 80th birthday last December came as a shock – still hard to accept that I am now in the venerable category – but grateful for many blessings – including reasonably good health, and that Jean still tries to keep me clean, tidy and overfed....
>
> Our reward of old age is enjoyment of memories – old friends and old times, like when I joined your camp on the Elk River end of the 1930 season. Shelley and I doing our morning dip in its ice-cold water, the trip down to the coast with you – the Davenport Hotel – Spokane, etc. When our time comes we can anticipate a wonderful reception committee within the Pearly Gates – FDM [Mulholland] – Shelley – CDS – Mc-Cannell – Gregg – Lex – Collins – CDO – Bill Hall – LGT – EWB – [all foresters who had died,] to name a few....
>
> Providence has dealt with me kindly – especially in things money can't buy. Am busy now with some memoirs of 60 years ago, and the days are not long enough.

After he returned to Victoria in 1984, George Newell accompanied Andrews on several trips. He remembers them as high-energy adventures,

George Newell at Akamina Pass.

rising early each day and covering a lot of country. Andrews was in his element driving on the back roads. He had an affinity for the people who lived in the rural areas they visited and had many friends living in remote places who always seemed happy to see him.

Newell recalls a 1985 trip to Prince Rupert when, en route, they detoured off the main highway to the Big Bar area where Andrews wanted to visit some friends. They reached the place early in the morning, so Andrews parked the van on the hill above the house until smoke rising from the chimney signalled that its residents had awoken. Even though it was morning Andrews' friends greeted their visitors with delight. After the visit, instead of returning to the highway, the two men crossed the Fraser River on the Big Bar ferry and took the gravel road along the west side of the river to Williams Lake where they visited Kris. During their 1989 trip to Vanderhoof they drove from Quesnel on gravel roads that followed the route of the Yukon Telegraph Trail.

Andrews and Newell made two trips in 1987. In May they travelled to Manitoba, to the areas related to Andrews' Soldiers of the Soil experience and his 1927 Manitoba Pulp survey, and to visit family. On the way there they went through Crowsnest Pass and down to Waterton. Andrews wanted to go to Akamina Pass and he also hoped to visit Monument 272, the place where the Alberta, BC and US boundaries all meet. But snow in the mountains prevented this from happening.

During the summer trip the men returned to the area. In his Christmas letter to Jack Aye Andrews described their adventure:

> You know how stubborn I am. I still wanted to look at Akamina, so latter
> August, George Newell and I went back again. We stopped at Cranbrook
> to check with the forestry people about getting there via the Flathead.
> They had no facilities to help, but said a helicopter outfit in Fernie was
> taking oil geologists in there. So next day we got a ride with them.
> Weather awful and couldn't land but got photo.
>
> We got back to Fernie mid day so drove on through to Waterton
> and Cameron Lake in time to park and climb up into the Pass before

sunset. Weather had sweetened sufficiently for the climb. The Pass has a bad reputation for grizzlies but they were busy somewhere else. You can imagine my memories up there, of driving your old Model T between our Akamina Camp and Waterton. We returned home via Banff and Field, and Jean was relieved that I was still not up in the Pass, having passed through the digestive tract of Mr. Grizzly! Gee, it is still beautiful country in spite of logging and fires below alpine barren.

Andrews and Newell also went to the park warden's office where the superintendent showed them some slate rocks that the wardens had found at Monument 272 with names carved on them. Andrews recognized the name of one of the international boundary surveyors.

A trip that Andrews and Newell took in 1990 was typical of their adventures. They left on May 8 and travelled through Washington where they visited the Grand Coulee dam. By May 10 the two men had reached Grand Forks and spent the weekend attending the annual meeting of the BC Historical Federation. Then they followed Highway 3 to Cranbrook to visit Andrews' relatives there. After staying overnight they travelled on Highway 93 south to Glacier National Park in Montana. They drove through the park in one day and camped that night in Waterton. From there they travelled across southern Alberta to Gull Lake, Saskatchewan, where they spent two days visiting Andrews' sister Leila and her family. Then they headed north, back to Alberta. On Sunday, May 20, the two men drove almost 800 kilometres to the Alberta Peace River region where they spent a day visiting some of Andrews' friends. The following day they travelled to the BC portion of the Peace River region, where they spent two days conversing with several friends and touring Kelly Lake. On the return trip they stopped to see people in the Fort St John and Hudson's Hope area. They camped that night at McLeod Lake, about two hours north of Prince George, then drove to Williams Lake, where they stayed at Kris's place. On Monday, May 28, they returned to Victoria, completing a 20-day loop trip that covered 5,045 kilometres.

Surveyor John Whittaker, who also travelled with Andrews a few times, remembers trips around northern BC that followed a similar pattern. Andrews continued his travels and visits to Atlin until the early 1990s.

Gerry Andrews received many awards for his years of service. Among them, the University of Victoria granted him an honorary doctorate in engineering in 1988. In 1990 he received the Order of British Columbia. The recipients were placed in alphabetical order, and to Andrews' amusement, he was beside famous rock musician Bryan Adams. And in 1991 Andrews was appointed a member of the Order of Canada.

Jean Andrews passed away in 2000. Gerry celebrated his centennial in 2003 by having a reception with friends and family from afar. He died in

2005 just a week before his 102nd birthday. In March 2011 the provincial government honoured him by naming Mount Gerry Andrews in the Flathead valley for him.

Gerry Andrews was interviewed several times during his retirement. One of the interviewers ended his program with quotation from Ecclesiastes 30:22: "the joyfulness of a man prolongeth his days". He could have also used the first part of the verse: "the gladness of the heart is the life of man". Andrews lived a long, happy life full of adventures.

Like one of his mentors, Frank Swannell, Andrews wanted to be an explorer and adventurer. Although he was not really an explorer, he travelled throughout British Columbia to many remote areas that few other people had seen. The Order of British Columbia citation describes Andrews as "truly one of British Columbia's great trail blazers".

During both his long, illustrious career and his retirement Andrews contributed to the history of British Columbia, especially through his work in aerial photogrammetry. In his 1968 government report Deputy Minister of Lands David Borthwick said, "Mr Andrews' pioneer work in the field of aerial surveying will always remain a credit to him and to his profession." Assisted by some exceptional mentors in his early career and by applying his talents and his persistence, Andrews proved the value of aerial photography to BC government officials in the 1930s. He then transferred his experience to the war effort, assisting the Allied commanders in the D–Day operations. After the war he returned to BC and expanded the knowledge and application of aerial photogrammetry to a wider audience in the BC government, to private industry and to the general public. As British Columbia's surveyor general he was involved in most of the major projects during the province's rapid economic growth of the 1950s and 1960s.

Andrews embraced new technologies that were developed during the 1950s and 1960s, adapted them to the province's specialized terrain and made BC one of Canada's leading provinces in modern survey methods. The Order of Canada citation for Andrews states: "An expert in photogrammetry, as Surveyor General of British Columbia and Boundaries Commissioner he contributed greatly to the geographic, economic and cultural development of his beloved province."

At the same time Andrews developed a close camaraderie with his employees and he inspired several of them to expand their professional careers. He had a wealth and variety of friendships that he maintained throughout his lifetime. His ability to make and keep friends is a comment made by everyone who knew him. Today there are many people who have fond memories of Andrews, both at his home in Victoria and during his travels to the remote areas of the province with his red sock and flask of hooch.

Acknowledgements

Many people assisted me in researching and writing this book. First and foremost I thank the Andrews family for making all of their father's material available. I appreciate Mary and Kris patiently answering all my questions and doing everything possible to assist. I especially appreciate Mary's willingness to provide a place for me to stay in Victoria after my son moved out of the city in the spring of 2011. This book could not have been finished on a tight schedule without Mary's assistance.

Thanks to the Association of BC Land Surveyors for making their research material available and for their enthusiasm for this book. And thanks to the surveyors who agreed to be interviewed, particularly Jon Magwood and John Whittaker.

George Newell spent an afternoon sharing his pleasant memories of the trips that he took with Gerry Andrews. I enjoyed listening to his account, and his insights were very helpful.

Eva Schindler and Stefan Himmer provided the transportation and co-ordinated our Labour Day weekend trip to the Flathead with Mary and Kris. We had a wonderful weekend hiking in the mountains and getting to the top of Trachyte Ridge for a close-up view of Mount Gerry Andrews. Harold Underhill assisted us on this hiking trip.

Geoff Swannell (Art's son) lent me his uncle Lorne's photo album and Flathead diary. It was nice to have the support of the Swannell family, for they were important in Andrews' work and personal life.

Thanks to all the people who provided material to assist with the research. And to Robert Allen for reading the surveying section and helping me with the material about Gerry Andrews.

Once again I would like to express my appreciation to my wife, Linda, for her support and the many ways she provided assistance. I would also like to also thank Gerry Truscott of the Royal BC Museum for publishing this book and producing it on time for the centennial of the BC Forest Service.

Sources Consulted

Andrews, Gerry S. 1931. Flathead Forest. Victoria: BC Forest Service,
Forest Surveys Division.

———. 1931. Tranquille Forest Survey. Victoria: BC Forest Service, Forest
Surveys Division.

———. 1932. Neskonlith Forest Survey. Victoria: BC Forest Service, Forest
Surveys Division.

———. 1932. Shuswap Forest Survey. Victoria: BC Forest Service, Forest
Surveys Division.

———. 1934. "Air Survey and Forestry: Developments in Germany".
Forestry Chronicle, June: 91-107.

———. 1936. "Tree Heights from Air Photographs". *Forestry Chronicle*,
June: 152-97.

———. 1940. "Notes on Interpretation of Vertical Air Photographs".
Forestry Chronicle, September: 201-15.

———. 1940. "The Small Scale Air Survey Camera". *Photogrammetric
Engineering* 6:2:91-97.

———. 1942. "Alaska Highway Survey in British Columbia". *The
Geographical Journal* C:1:5-22.

———. 1944. "Check Soundings for WV Profiles on the Normandy
Coast", November 1944. Supplement No. 1 to *Wave Velocity Soundings
from Air Photographs*.

———. 1945. "Rectifying the Extra-axial Exposure Distortion of Wide-
angle Air Survey Lenses". *Canadian Army Overseas*, March 1945.

———. 1946. "Survey and Mapping in Various War Theatres: Report of
a Special Mission in 1945 for the Canadian Army Overseas." Ottawa:
Geographical Section, General Staff, Department of National Defence.

———. 1948. "Air Survey and Photogrammetry in British Columbia".
Photogrammetric Engineering, March: 134-53.

Andrews, Gerry S. 1954. "Surveys and Mapping in British Columbia Resources Development." Paper for the British Columbia Natural Resources Conference.

———. 1958. Surveys and Mapping Requirements for Development Planning Lower Mekong River Southeast Asia. Victoria: Office of the Surveyor-General.

———. 1960. "Some Statutory Aspects in Cadastral Use of Photogrammetry". *Canadian Surveyor* 15:309-16.

———. 1962. "British Columbia's Major River Basins". Paper for the 14th British Columbia Natural Resources Conference.

———. 1964. "An Integrated Survey System for British Columbia". *Canadian Surveyor* 17:2:119-28.

———. 1969. "The Surveyor's Role in Resource Development". American Congress on Surveying and Mapping, Washington, DC, pp. 673-78.

———. 1970. *Administration of Surveys and Mapping in Canada, 1968.* Ottawa: National Advisory Committee on Control Surveys and Mapping.

———. 1973. "British Columbia's Air Survey Story". *BC Historical News,* November: 18-27.

———. 1974. "British Columbia's Air Survey Story, Part 2". *BC Historical News,* February: 16-26.

———. 1974. "British Columbia's Air Survey Story, Part 3". *BC Historical News,* April: 15-21.

———. 1975. "A Traverse of East Kootenay Survey History". *BC Historical News,* February: 13-26.

———. 1976. "Edouard Gaston D. Deville". *BC Historical News,* April: 17-23.

———. 1978. "Professional Land Surveyors of British Columbia". Cumulative Nominal Roll, 4th edition. Victoria: BC Land Surveyors.

———. 1979. "A Small Profession in a Large Land: Pioneer Land Surveyors of British Columbia". *BC Historical News,* Winter: 7-18.

———. 1979. "The Bell-Irving Land Surveyors in British Columbia". *BC Historical News,* Summer: 11-15.

———. 1982. "First World War Aviators Paved Way for BC Surveys". Victoria *Times-Colonist,* August 29, pp. 4-5.

———. 1982. "Photo-flying in the BC Public Service". Victoria *Times-Colonist,* September 5, pp. 8-9.

———. 1985. *Metis Outpost.* Victoria: Gerry Smedley Andrews (self published).

———. 1987. "Beyond those Rugged Mountains". *BC Historical News,* Fall: 12-14.

———. 1988. "Beyond those Rugged Mountains". *Alberta History,* Winter: 11-18.

———. 1988. "What the War did to Me". Unpublished article.

———. 1989. "Major Richard Charles Farrow". *Link*, March: 3-6.

———. 1989. "The Makings of a Surveyor-General". *Link*, July: 12-20.

———. 1989. "My Bear". *The Victoria Naturalist* 46.3:5-6.

———. 1989. "Reminiscences of a Soldier of the Soil". *Manitoba History*, Spring: 26-30.

———. 1991. "The Flathead Forest Survey – 1930". Unpublished article.

———. 1995. *Big Bar Country*. Unpublished booklet.

Andrews, Gerry, and Lyle Trorey. 1933. "The Use of Aerial Photographs in Forest Surveying". *Forestry Chronicle*, December: 33-59.

Andrews, Mary, and Doreen Hunter. 2003. *A Man and his Century: Gerry Smedley Andrews.* Victoria: Mary Andrews and Doreen Hunter (self published).

BC Archives. 1987. "A Proud Past." Gerald S. Andrews interview, March 9. BC Archives, AAAA 5271.

BC Historical Federation. 2011. "Flathead Mountain Named in Honour of Legendary Surveyor, Gerald Andrews". BC Historical Federation Newsletter, June, p. 3.

BC. Forest Service. 1950. "Flying Surveyors". BC Archives, AAAA1887.

British Columbia, Government of. 1938 to 1940, 1947 to 1969. Sessional Papers.

Burch, Gerry and Parminter, John. 2008. *Frederick Davison Mulholland P.Eng., BCRF: the Father of Sustained Yield Forestry in British Columbia.* Victoria: Forest History Association of BC.

Canada, Government of. 1955. Report of the Commission Appointed to Delimit the Boundary between the Provinces of Alberta and British Columbia: Part 4, 1950 to 1953, Latitude 57°26'40"25 Northerly. Ottawa: Office of the Surveyor General.

———. 1959. Memorandum from Secretary of State for External Affairs to Cabinet: Document No. 43-59, Ottawa, February 4, 1959. Ottawa: Foreign Affairs and International Trade Canada, www.international.gc.ca.

———. 1966. Report of the Commission Appointed to Delimit the Boundary between the Province of British Columbia and the Yukon and Northwest Territories. Ottawa: British Columbia–Yukon–Northwest Territories Boundary Commission.

Chadwick, Vivienne. 1963. "D-Day Beaches His Business". Victoria *Daily Colonist*, March 31, pp. 4-5.

Cooper, Richard W. 1985. "He Introduced Aerial Mapping to BC" Victoria *Times-Colonist*, February 3, p. 12.

Cornwall, George L. 1999. *One Damn Swamp After Another!* Kanata, Ontario: Sandwell Books.

Ellis, Grant. 1978. "Taking Inventory: from Saddles to Satellites". *ForesTalk*, Fall, 11–15.

Gidney, Norman. 2005. "Photos Cleared Way for D-Day".Victoria *Times-Colonist*, December 16, p. B11.

Gordon, Katherine. 2006. *Made to Measure: A History of Land Surveying in British Columbia*.Winlaw, BC: Sono Nis Press.

Gould, Jan. 1986. "A backwoods education for the teacher".Victoria *Times-Colonist*, March 9, pp. M3–4.

———. 1986. "Summer Jobs Laid Foundation for Career".Victoria *Times-Colonist*, March 16, pp. M3–4.

———. 1986. "War Sent Andrews Back Across the Atlantic".Victoria *Times-Colonist*, March 23, pp. M3–4.

Hodgins, H.J. 1931. Elk Forest.Victoria: BC Forest Service, Forest Service Division.

Hore, Major W.H. 1944. "Memorandum on Photography for Beach Gradient Determination". The War Office, London, UK.

Kelley, Frank. 1947. "Shooting BC from the Air." Vancouver *Daily Province*, March 8, pp. 1,7.

Lillard, Charles. 1979. "Gerry Andrews". *ForesTalk,* Winter: 28–29.

Mahood, Ian, and Ken Drushka. 1990. *Three Men and a Forester*. Madeira Park, BC: Harbour Publishing.

Mulholland, F.D. n.d. Instructions for Forest Surveys.Victoria: Province of British Columbia, Department of Lands, Forest Branch.

Oxford University Forest Society. 1933. "The Jura, Switzerland and the Black Forest", *Journal of the Oxford University Forestry Society*, Michaelmas Term, pp. 12–14.

Oxford University Forest Society. 1933. "The Working Plan Tour, 1933" *Journal of the Oxford University Forestry Society*, Michaelmas Term, pp. 6–9.

Perth *Courier*. 1918. "*Lack of Food Threatens Battle Lines*" Perth (Ontario) *Courier*, March 29.

Rogers Cable. n.d. Gerald S. Andrews interviews. BC Archives, AAAA1631.

Schultz, C.D. 1934. The Nimpkish Forest.Victoria: BC Forest Service, Forest Survey Division.

Sherwood, Jay. 2010. *Return to Northern British Columbia: A Photojournal of Frank Swannell, 1929–39*.Victoria: Royal BC Museum.

Stewart, N.C. 1954. "Mapping Alternative Routes, Hazelton to Atlin". *BC Professional Engineer*, December: 17–19.

Swannell, Frank. n.d. Records. BC Archives MS-392 files.

Swannell, Lorne. 1930. Diary (unpublished). Swannell family collection.

Thiffault, Pierre. "Ellwood Wilson, Forester and Visionary: An Introspective Study". *Canadian Aviation Historical Society* 42:4: 124–131, 154–155.

Thomson, Don W. n.d. Window on the Third World: The Role of the Federal Surveys and Mapping Branch in Canadian Aid Overseas. Ottawa: Energy, Mines and Resources Canada.

Times-Colonist, Victoria. 1950. "British Columbia Air Mapped in Five Years". Victoria *Times-Colonist*, November 19, p. 4.

———. 1967. "Key marker unveiled". Victoria *Times-Colonist*, June 22, p. 26.

Index

Jay Sherwood

From 1979 to 1986, Jay Sherwood lived and taught school in Vanderhoof. There, he learned about the legendary land surveyor, Frank Swannell through his involvement with the local history society. A former surveyor, himself, Sherwood embarked on a study of Swannell that would result in the publication of three books about him. While researching Swannell's work, Sherwood uncovered a memoir written by Bob White (1902–85), a cowboy and packer who worked on the Bedaux Expedition, in which Swannell played a small part. He transcribed White's account, edited it and wrote a contextual framework for the story.

Jay Sherwood recently retired from his teaching position in Vancouver to concentrate on his historical research projects. His interest in another important and influential land surveyor, Gerry Andrews, resulted in this book. And he's now working on a book about surveyor George Milligan and the explorer and scoundrel E.B. Hart. "In the Shadow of the Great War: The Milligan and Hart Expeditions to Northeastern BC, 1913–14", will be published by the Royal BC Museum in 2013.

Other Books by Jay Sherwood:

Surveying Northern British Columbia: A Photojournal of Frank Swannell (Caitlin Press 2005)

Surveying Central British Columbia: A Photojournal of Frank Swannell, 1920–1928 (Royal BC Museum 2007)

Bannock and Beans: A Cowboy's Account of the Bedaux Expedition (Royal BC Museum 2009)

Return to Northern British Columbia: A Photojournal of Frank Swannell, 1929–1939 (Royal BC Museum 2010)

The Royal BC Museum

British Columbia is a big land with a unique history. As the province's museum and archives, the Royal BC Museum captures British Columbia's story and shares it with the world. It does so by collecting, preserving and interpreting millions of artifacts, specimens and documents of provincial significance, and by producing publications, exhibitions and public programs that bring the past to life in exciting, innovative and personal ways. The Royal BC Museum helps to explain what it means to be British Columbian and to define the role this province plays in the world.

The Royal BC Museum administers a unique cultural precinct in the heart of British Columbia's capital city. This site incorporates the Royal BC Museum (est. 1886), the BC Archives (est. 1894), the Netherlands Centennial Carillon, Helmcken House, St Ann's Schoolhouse and Thunderbird Park, which is home to Wawaditła (Mungo Martin House).

Although its buildings are located in Victoria, the Royal BC Museum has a mandate to serve all citizens of the province, wherever they live. It meets this mandate by: conducting and supporting field research; lending artifacts, specimens and documents to other institutions; publishing books (like this one) about BC's history and environment; producing travelling exhibitions; delivering a variety of services by phone, fax, mail and e-mail; and providing a vast array of information on its website about all of its collections and holdings.

From its inception 125 years ago, the Royal BC Museum has been led by people who care passionately about this province and work to fulfil its mission to preserve and share the story of British Columbia.

**Find out more about the Royal BC Museum at
www.royalbcmuseum.bc.ca**